STATISTICS IN PLAIN ENGLISH

HARVEY J. BRIGHTMAN
Georgia State University

Duxbury Press
An Imprint of Wadsworth Publishing Company
Belmont, California

Duxbury Press
An Imprint of Brooks/Cole Publishing Company
Pacific Grove, California

ISBN: 0-538-13210-8
Library of Congress Catalog Card Number: 85-62661

12 13 14 - 01

Printed in the United States of America

Preface

The subject of this book is statistics, but it's not like any other statistics text on campus. This is a self-teaching text for students who need it to use on their own. It is written for anyone who has difficulty with the way statistics is usually presented, or who is simply intimidated by the subject. It can also serve as a friendly introduction for those who have not yet had a course in statistics, or as a refresher for those who had one some time ago.

I've tried to keep a light, friendly tone in the writing, but the pedagogical approach is more serious than it looks. It's based on the principles underlying mastery learning—the achievement of a true working mastery of a subject. Cognitive psychologists tell us there are two conditions necessary for mastery learning to occur. First, the subject matter must be presented and organized along mental routes with which students are familiar. Second, students must be active learners and must discover the important concepts underlying the material.

One leading cognitive psychologist, Jerome Bruner, believes that students must learn to articulate ideas and concepts in ways that are familiar to them. If they cannot do that correctly, they haven't yet mastered the material. Formulas may confuse students who are not mathematicians, and may actually obscure the statistical concepts. This mismatching between text presentation and student cognitive orientation is one reason why students fail to achieve mastery in statistics. If mastery is to occur, they must learn to translate mathematical concepts into plain English.

The second principle of mastery learning—discovery learning—is based upon the idea that students can retain, utilize, and modify what they learn if they can be led through a process of self-discovery. Students must be given the opportunity to arrive at their own conclusions about the subject matter before being told the hows and whys. If they do this, they will be able to go beyond what they know.

Incorporated throughout *Statistics in Plain English* are three unusual features which are derived from the principles of mastery learning. At the beginning of most sections within each chapter is a list of behavioral objectives. Not enough books provide behavioral objectives, and those that do often focus on the content of the material rather than on what the student will be able to do with it. I have found the Bloom, et al, taxonomy of cognitive objectives very useful in designing my courses and have incorporated it into the book. It provides students with a specific set of observable skills that they should have mastered by the end of each unit. So students will know what they should be able to do with the statistical concepts and ideas contained within each chapter.

An important concept in effective teaching is that students must be active learners, not passive readers. Therefore, throughout the book I ask questions and ask readers to write down their thoughts in the space provided. In this way the readers become joint authors as they write their own translations of the material that is presented. If students are to master the material, they must be given a chance to discover the concepts before being told what they are. I have tried to construct these questions, and the answers which immediately follow, in such a way that even if the student fails to completely discover the concept, he or she will learn from the effort and feel good about it.

Finally, at the end of most sections there is a series of problems and short essay questions. The complete solutions to all problems are in the back of the book. This book has a larger percentage of short essay questions than many books because my philosophy is, "if you can't say it in plain English, you don't know it."

These ideas are based on 14 years of teaching statistics to undergraduate and graduate business students. I have attempted to anticipate problems that students often experience as they try to learn the whys as well as the hows. *Statistics in Plain English* focuses on what students have told me they consider the five hardest topics in statistics. The book presumes no background in calculus, and minimizes the use of the mathematical language. I generally use the active voice, move from the simple to the complex and the familiar to the unfamiliar, and encourage readers to put into their own words what they are reading. I also use extended examples throughout a chapter (rather than many disconnected examples) because educational theory suggests that this improves learning.

Many of the educational innovations in this book stem from the seminar on university teaching that I have offered to our doctoral students at Georgia State University for the past six years. I have exported the ideas of mastery learning and effective teaching to over 200 Ph.D. candidates and faculty throughout the United States and Europe. I hope these concepts are as successful in this book as they have been in the classroom.

I would like to dedicate this book to my wife Arlene, the mean of my life, and my daughters, Ellen and Pam, the variances.

Harvey J. Brightman

Contents

1.
Introduction and Descriptive Statistics

1:1 A NOTE TO THE READER

Formulas are not the everyday language of most business students and managers. Yet most statistics books present the material in a mathematical manner. I believe that statistics should be presented in a way that students or managers, not teachers, learn best. That is the purpose of this book.

Instead of relying heavily on formulas, this text uses pictures and plain English. Visual images are used to show relationships and make rough estimates; only then are formulas used to compute the correct answer. Even then I avoid the use of Greek symbols for the mathematical notation to keep it as familiar as possible. Instead, I try to explain the logic and common sense behind the formulas in plain English. My objective in this text is to help you to correctly translate statistical ideas into your own words.

You need to be an active learner while reading this book. To help you put statistics in your own words, I'll occasionally ask you to think through an idea and to write your explanation in the space provided. When answering these questions, imagine yourself talking to a friend who doesn't know much about statistics. In this way, you will be able to explain statistical ideas without resorting to formulas and statistical jargon. When you can do this, you have mastered the material.

At the beginning of most subsections you will find a set of behavioral objectives. They describe the skills you should have after completing the subsection. Unless you can accomplish all the objectives, you should not proceed to the next unit. Also provided are some exercises that go beyond just plugging numbers into a formula.

They are based upon the behavioral objectives at the beginning of each subsection. You'll find the answers to all the problems in the back of the book.

This book covers the five hardest topics in statistics. (In Chapter 1, some relatively easy introductory material is presented on descriptive statistics.) Chapter 2 covers basic probability, statistical independence, and Bayes' Theorem without using the complicated formula associated with this theorem. Discrete and continuous probability distributions are covered in Chapter 3 with a minimal number of formulas. The focus there is on how to determine when to use two of the most common probability distributions—the binomial and the normal distributions—to solve a problem. This is important because I can never recall a problem walking up to me and saying, "I'm a binomial problem—solve me." An alternative method for estimating probabilities—simulation—is also presented in Chapter 3. Chapter 4 explains what an inductive inference is and how to make valid generalizations from small samples (similar to what the Gallup organization does). Chapter 5 explains how to design experiments and analyze the data using inductive inference. You'll need only one formula and a number of visual techniques to analyze experimental data. Regression analysis, or looking for relationships between variables, is covered in Chapter 6. A blend of visual, verbal, and mathematical approaches are presented there in the hope that at least one approach will appeal to you.

The material covered in this text is traditional but the presentation is not. Hopefully this approach will eliminate the fear and mystery of statistics. If you don't let the math get in your way, you'll find that statistics is not sadistics. The key is to translate what you are learning into your own words.

1:2 INTRODUCTION

Statistics involves collecting data, organizing and summarizing it, and analyzing the data to make decisions. For example: the Gallup organization wants to make a prediction on an upcoming election; a city wishes to know if it is eligible for a federally funded program which is based on a maximum average income level within a city; a researcher wants to know if increasing the dosage of vitamin C will reduce the number of colds during the winter. These examples are instances of statistics at work and each involves collecting, organizing, and analyzing data. Let's look at these three areas of statistics in more detail.

1:2:1 Collecting Data

In the 1984 U.S. presidential election, most of the polling organizations predicted that President Reagan would win by a 16 to 22 percent margin. He won by 19 percent. How can polling results be so accurate when less than 2,000 out of the 90,000,000 voters are interviewed? It's the way in which the organizations collect their data. Polling organizations have developed sampling procedures that ensure that they obtain a representative sample. The sex, race, income level, religion, geographical locations, etc., of the sample must be similar to the population of voters. (This procedure is explained in Chapter 4.) Once the sample is selected, the polling organizations must phrase their questions properly. For example, "If the election were held today, who would you vote for?" is better than "Which candidate do you like?" People may like a candidate and still not vote for him or her.

How can a city determine its average income level to see if it qualifies for a federal program? Every ten years the government takes a complete census of the population. However, during the ten years in between there is no way to know with absolute certainty what the average income is. The city agency must take a representative sample, determine its income level, and make an educated guess about the average income level of the city. We call an educated guess an **inductive inference** (more on that in Chapter 4).

Determining the income level for each family or unit in the sample may not be easy. The city agency may not have access to Internal Revenue Service files. Thus, the agency may need to ask each family to report their earnings. This may lead to errors. While the city agency will attempt to double-check the data, it may be difficult.

In the vitamin C study, the researcher will have to select subjects who are reasonably healthy. If some are ill at the beginning of the study, the researcher may draw incorrect conclusions. Each subject must also be given the proper dosage of vitamin C each day. The researcher will then have to record the number of colds during the winter season for each subject. The researcher will also have to ensure that the subjects do not eat foods such as oranges which contain vitamin C. If that happens, a subject who is supposed to get 1,000 mg. of vitamin C may actually be getting a higher dosage. This will make the results meaningless. How to conduct formal experiments will be discussed in Chapters 5 and 6.

1:2:2 Organizing and Summarizing Data

Frequently we need to summarize the raw data from a study. In the Gallup study, what percentage within the sample said that they would vote for President Reagan? It's the **sample percentage**, which is simply the number who favored President Reagan divided by the total number in the sample.

In the city eligibility study, we need to compute the average income level for the sample. Do we expect each family in the sample to have the same, exact income level? No. Therefore, we must also compute a measure of how different the incomes were in the sample. This is called the **standard deviation**. Because there may be thousands of family income levels in our study, we may want to summarize them into one compact picture called a **histogram** (more on that later).

In the vitamin C study, the data would be summarized as shown in the following table.

Summary of Vitamin C Sample Data

Amount of Vitamin C per Day	Average Number of Colds
0 mg.	4.5
1,000 mg.	3.0
2,000 mg.	2.5
3,000 mg.	2.0
4,000 mg.	0.5

1:2:3 Analyzing Data

Analyzing data calls for making educated guesses. Knowing what percentage in the *sample* supports a candidate is not enough. We want to know what percentage of *all* 90,000,000 voters will support the candidate. After all, that is the purpose of conducting polls. We'll use the sample percentage to estimate the percentage of the population who will support the candidate. To do this we'll use an inductive inference procedure called a **confidence interval**. Obviously we can never be 100 percent confident that our estimate is accurate since we have only taken a very small sample from the population of all eligible voters. However, you'll be surprised how close your estimates will be.

Looking for a relationship between the amount of vitamin C and the number of colds also calls for making an educated guess. Based upon a small sample, the researcher wants to generalize the findings.

Here the inductive inference procedure to analyze the data is **regression analysis** (presented in Chapter 6).

Probability and statistics are more than mere data analysis. You must never forget where and how the data were collected. Several centuries ago Sir Josiah Stamp said it this way:

> Public agencies are very keen on amassing statistics—they collect them, raise them to the *n*th power, take the cube root and prepare wonderful diagrams. What we must never forget is that every one of these figures comes in the first instance from a village watchman, who just puts down what he damn pleases.

With that quote in mind, let's get started. The first topic is descriptive statistics and histograms—the art and science of transforming a mass of confusing data into meaningful information.

1:3 CENTRAL TENDENCY

By the end of this unit you should be able to:

1. Explain the need for summarizing data into a histogram.
2. Compute the mean for a set of raw data.
3. *Estimate*, using a *graphical* approach, the mean of a histogram.
4. Compute the mean of a histogram.
5. Determine when to use the median or mean.
6. Determine the median for a set of raw data.
7. Compute, using a *graphical* approach, the median of a histogram.

1:3:1 The Mean or Average for Raw Data

Imagine that we have taken the pulse rates of ten well-conditioned athletes. The data, which have already been rank-ordered, are shown below. What is the mean or average for the ten numbers?

Pulse Rates of Ten Well-Conditioned Athletes (beats per minute)

31	38
33	39
36	41
36	44
37	47

The **mean** or average of a set of raw data is the sum of all the observations divided by the number of observations.

The average pulse rate (commonly known as *x*-bar) for the ten athletes is

$$x\text{-bar} = (31 + 33 + 36 + 36 + 37 + 38 + 39 + 41 + 44 + 47) \div 10$$

$$= 38.2 \text{ Beats per Minute}$$

1:3:2 The Mean for Tabled Data

We construct a histogram to make a large mass (or is it mess?) of data meaningful to others. Remember the example of several thousands of families in the city eligibility study. We would certainly want to organize and summarize that data before presenting the results to the city council. How do you think the city council will react if we hand them ten pages of income level data? They won't be able to decipher it. By organizing the raw data into one histogram, the confusing data can be transformed into meaningful information.

Let's draw a histogram for the pulse rate data set. The starting point is a **frequency table**. The table shows the number of athletes with various pulse rates. The variable of interest—pulse rate in beats per minute—must be subdivided into a number of classes. The classes must be constructed so that they (1) do not overlap and (2) include all the observations. If they overlap, an observation may show up in two classes and the number of observations in the histogram will exceed the original sample size. If the classes do not include all the observations, then the number of observations in the histogram will be smaller than in the original sample. Either way, the histogram will not correctly portray the data. To avoid false impressions, you should make the class widths equal size.

Look at the frequency table constructed below.

Frequency Table

Pulse Rate Classes	Frequency (number of athletes)
30 to 34.99	2
35 to 39.99	5
40 to 44.99	2
45 to 49.99	1
	10

Note that the classes do not overlap and do include all the athletes. Each class also has the same width of 4.99 or 5. Now we can define a histogram and then draw one.

A **histogram** is a picture or diagram that shows the number of observations that fall into each class; this is shown as a block in the histogram. The height of each block represents the number of observations within the class.

Figure 1-1 Pulse Rate Histogram

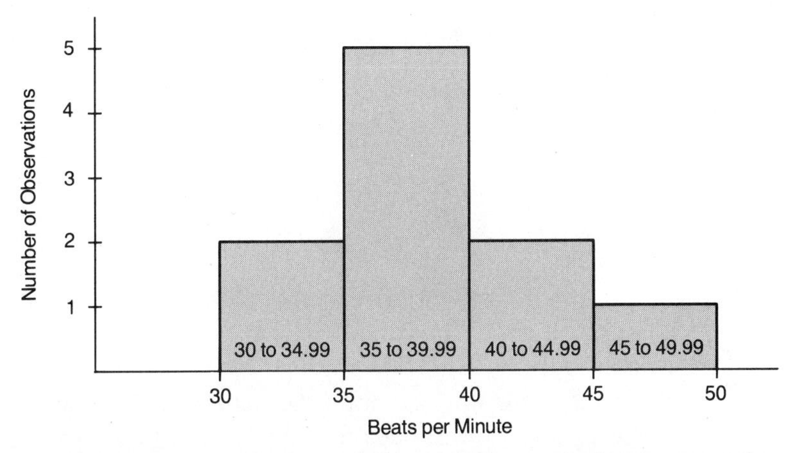

There is a quick way to estimate the mean of a histogram. The mean is at the *balance point* of the histogram. Think of the observations as weights and the x-axis of the histogram as a wooden board. Each class has a stack of weights, one for each observation. Below the wooden board is a steel rod. Move the rod back and forth. Where the board balances is the mean of the histogram.

Figure 1-2 Balance Point of the Pulse Rate Histogram

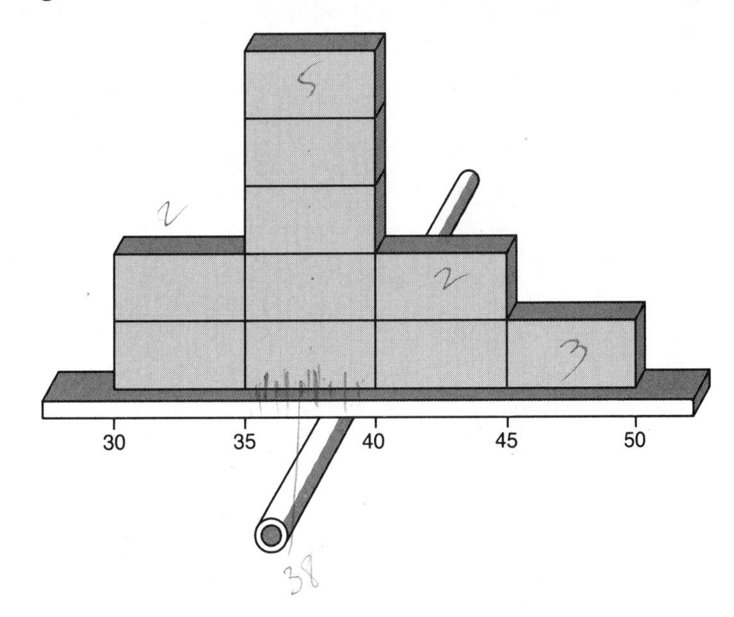

Let's apply this idea to the pulse rate histogram. At what pulse rate would the histogram balance? It wouldn't balance at 35 beats per minute because there would be too much weight (too many observations) to the right and the histogram would tilt down to the right. It wouldn't balance at 45 beats per minute because there would be too much weight to the left and the histogram would tilt down to the left. At *roughly* what pulse rate would the histogram balance? Please do this before reading on.

I hope your estimate of the mean of the histogram was in the upper thirties. You can't do any better than that. After all, the balance point idea only provides a rough estimate of the mean.

If you wish to obtain the exact value of the average for the histogram you will need to compute it. The expression for computing the mean of a histogram is actually derived from the balance point idea.

Take the number of observations in the first class and place the weight at the center or midpoint of the class. The midpoint of the first class is $(30 + 34.99) \div 2$ or 32.5. Now do it for all the classes. If you close your eyes, you should be able to see a two-pound weight hanging from the midpoint of the first class, a five-pound weight hanging from the midpoint of the second class, a two-pound weight hanging from the midpoint of the third class, and a one-pound weight hanging from the midpoint of the fourth class.

Now, in order to compute the *exact* balance point (the mean), you must multiply the weights by their respective midpoints, sum them, and divide by the total number of *observations*.

$$x\text{-bar} = [(32.5 \times 2) + (37.5 \times 5) + (42.5 \times 2) + (47.5 \times 1)] \div 10$$

$$= 38.5 \text{ beats per minute}$$

Hopefully, the computed value is close to your quick estimate based upon the balance point idea. Some additional histograms are provided at the end of the unit for further practice. Note that the means for the raw data and the histogram are not exactly the same (they are close, but not identical), although they are based upon the same data set. Take a minute and jot down why you think they are not the same. Don't read ahead.

The mean for raw data is computed using the values of all the observations. The mean of the histogram used only the midpoints of the classes and the number of observations in each class. For the convenience of summarizing the data we pay a price—we give up some information and lose some precision.

The mean of a histogram depends upon the width and the number of classes we choose to display the raw data. Although you have been given some general guidelines for building histograms, there is no one best way to display the data. If you change the number or width of the classes, this will change the midpoints and the number of observations within each class. From the expression you should be able to see that the value of the mean might change slightly. Try it. Redefine the classes into widths of six (the present widths are five) starting at 30 beats per minute. Redraw the histogram and compute its mean. The value of the mean will change slightly.

If you defined three classes including the values 30 to 35.99, 36 to 41.99, and 42 to 47.99, you should have computed a mean of 39. Grouping the data into a different set of classes results in a different computed value for the mean.

1:3:3 The Median for Raw Data

Suppose a company had ten employees and the average or mean salary was $100,000 per year. Would you like to work for such a company? Your inclination is probably to say, "Where do I sign up?" But think about it for a minute. If the average salary is $100,000 that does not guarantee that you'll be getting big bucks. Suppose that nine employees earn $10,000 per year and the big boss earns $910,000 per year. The average for the ten people is $100,000. The moral to this story is that the average may sometimes deceive us. We need another measure of central tendency.

The *median* is the value in the middle when all the observations are ranked in ascending order.

To determine the median for a raw data set you should:

1. Rank-order the data in ascending order.
2. Select the value that subdivides the ranked data set so that one half of the observations lie above it and one half below it. That observation is the median which is also known as the 50th percentile.

Now we will determine the median for the pulse rate data. The data are shown below ranked in ascending order.

31, 33, 36, 36, 37, 38, 39, 41, 44, 47

The median can't be 37 since there would be four numbers below it (31, 33, 36, 36) and five numbers above it (38, 39, 41, 44, 47). The median must be between 37 and 38; that is, somewhere between Observations 5 and 6. Arbitrarily we say that the median is halfway between the two values, or 37.5

In this example the mean and median were relatively close. In the salary example the mean was $100,000 but the median was only $10,000 (check this). Why do you think the median may not always be close to the mean even though they both are measures of central tendency? Ask yourself, what is the major difference between how we compute the mean and how we determine the median. That's where the answer lies.

When we determine the median we are not concerned with the actual values of the observations that lie above and below the median. We are only concerned that the *number* of observations are the same. When we compute the mean we sum the *actual* values of the observations and divide by the sample size. Thus, if you have a few observations that are either much greater or much smaller than the other observations (these odd observations are called **outliers**), the median may be very far away from the mean.

The median is often used to summarize demographic data. We often hear of the median family income level within a city. The reason the median is used is that statisticians don't want a few very rich families to affect their measure of central tendency. If a few wealthy families live in a small community, the mean income level is shifted upward but the median is not affected. Thus, the median provides a more accurate picture of the income levels in the city. The mean can be misleading when there are a few outlier data values. Remember the salary example! That's when it is best to use the median.

If there are no outliers, the mean and the median will usually be close to one another. But be careful of those who might try to trick you by providing you with an inappropriate measure of central tendency. Your best bet would be to ask for both measures of central tendency.

1:3:4 The Median for Tabled Data

A graphical way to determine the median for tabled data in a histogram is presented next. Let's compute the median for the pulse rate data histogram. The histogram is redrawn below.

Figure 1-3 Pulse Rate Histogram

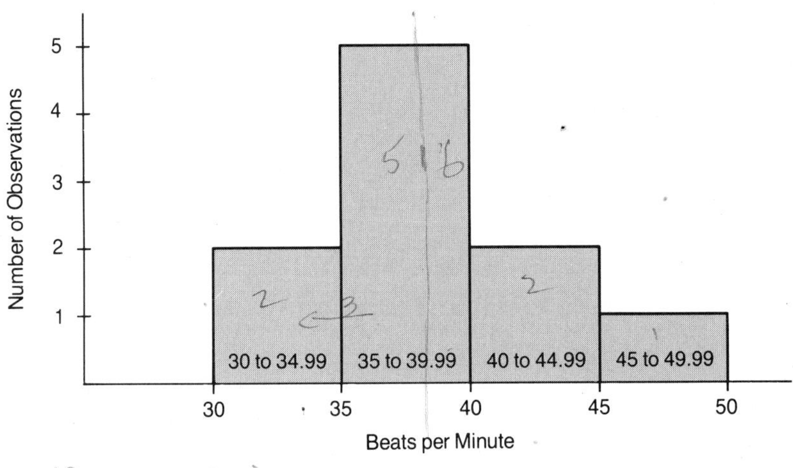

Here's the procedure:

1. Determine which observation (or between which two observations) is the median value.

As there are ten observations, the median is halfway between Observations 5 and 6. Five observations lie above and below the median.

2. Determine the class that contains the median.

Observations 5 and 6 lie in the second class—the one that goes from 35 to 39.99 beats per minute. How do we know this? The first two observations fall into the first class; the next five observations fall into the second class. These include Observations 3, 4, 5, 6, and 7. Thus, Observations 5 and 6 fall into this class.

3. Space the observations out evenly within the class. To do this divide the class that contains the median into subintervals of *equal length,* one for each observation in the class.

In our example the number of observations in the class that contains the median is five. Therefore, subdivide the interval from 35 to 39.99 beats per minute into five equal subintervals. This is shown below.

Figure 1-4 Line Graph Dividing the Second Class into Subintervals

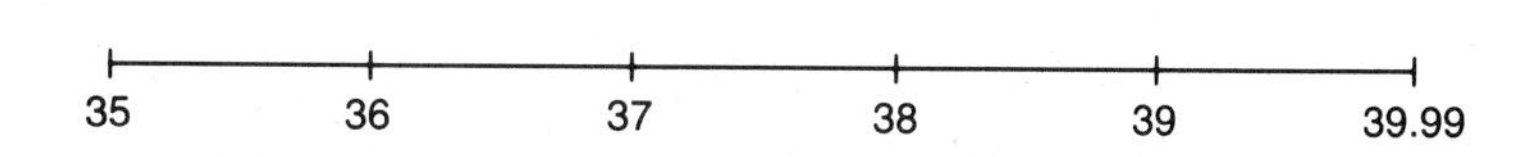

Determine the widths of the subintervals by dividing the width of the interval (39.99 − 35 = 4.99 or 5) by the number of observations in the interval (5). Thus, each subinterval is equal to one beat per minute.

4. Place each of the observations in the class that contains the median observation(s) at the midpoint or center of each subinterval. Place the first observation in the class at the midpoint of the first subinterval and continue until all observations have been placed on the line graph. Locate the median observation(s) and read the corresponding value from the line graph.

Figure 1-5 Line Graph for Determining the Median

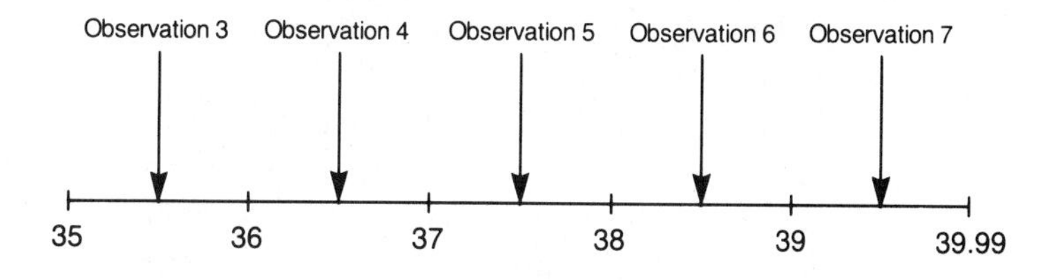

The median for ten observations lies halfway between Observations 5 and 6. According to the line graph, halfway between these two observations is 38. Thus, the median is 38 beats per minute. The four-step procedure works every time.

While the mean of a histogram is its balance point, the median splits the histogram so that half the observations lie above it and half below it. Here are two more histograms. For which histogram will the mean be larger than the median? Take a few minutes and write your answer below.

Figure 1-6 Two Histograms—For Which Is the Mean Larger Than the Median?

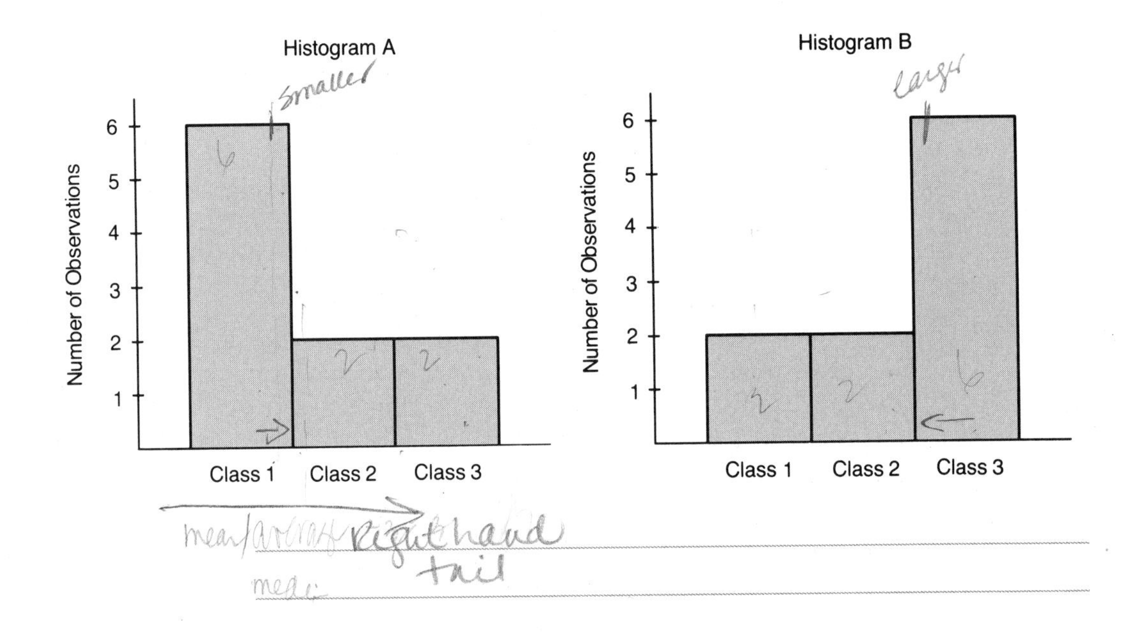

Hopefully you concluded that the mean will be larger than the median for Histogram A and the mean will be smaller than the median for Histogram B.

Histogram A is called a long right-hand tailed histogram. The mean will be further to the right (larger) than the median because the observations in Class 3 will cause the balance point to shift towards the right. The median will lie within Class 1 between Observations 5 and 6, but the mean will probably lie somewhere in the beginning of Class 2. Use the balance point idea to check this thinking.

Histogram B is called a long left-hand tailed histogram. The mean will probably lie somewhere in Class 2. The median will lie in Class 3 between Observations 5 and 6.

Thus, for a long right-hand tailed histogram, the mean will be larger than the median. For a long left-hand tailed histogram, the mean will be smaller than the median. Under what conditions would you expect the mean and median to be about the same?

You are correct if you said for a histogram that is neither right-hand nor left-hand tailed. These are called near-symmetric histograms.

Please review the objectives for this unit. If after completing the exercise set below you haven't mastered the objectives, reread the material.

Exercise Set for 1:3

1. Find the mean and median for each of the following data sets:

   1, 5, 6

   −1, 3, 4, 89

   1, 3, 789, 9,000, 245,678

2. At an urban university there are 7,000 undergraduates who are 18–23 years old, 2,000 undergraduates who are 24–29 years old, 1,000 undergraduates who are 30–35 years old, and 1,000 undergraduates who are 36 years old or older. *Without doing any math,* explain in plain English why the mean will be larger than the median.

3. There once was a statistician who couldn't swim. He came to a river but before he crossed it he asked the farmer what the mean depth of the water was. The farmer replied that the average depth was only two feet. The statistician smiled (for he was six feet tall) and began to cross the river. He drowned. What had he forgotten about the mean?

4. A group of 12 meteorology students were asked to guess at the average annual temperature in Reno, Nevada. Their guesses are shown in the histogram on the next page. (What would you guess? I'll give you the actual value with the solutions.)
   a. Using the balance point idea, estimate the mean of the histogram.
   b. Compute the mean and median of the histogram.
   c. What conclusion about the mean and median can you draw for a symmetric histogram?

Figure 1-7 Reno Nevada Average Annual Temperature Estimates

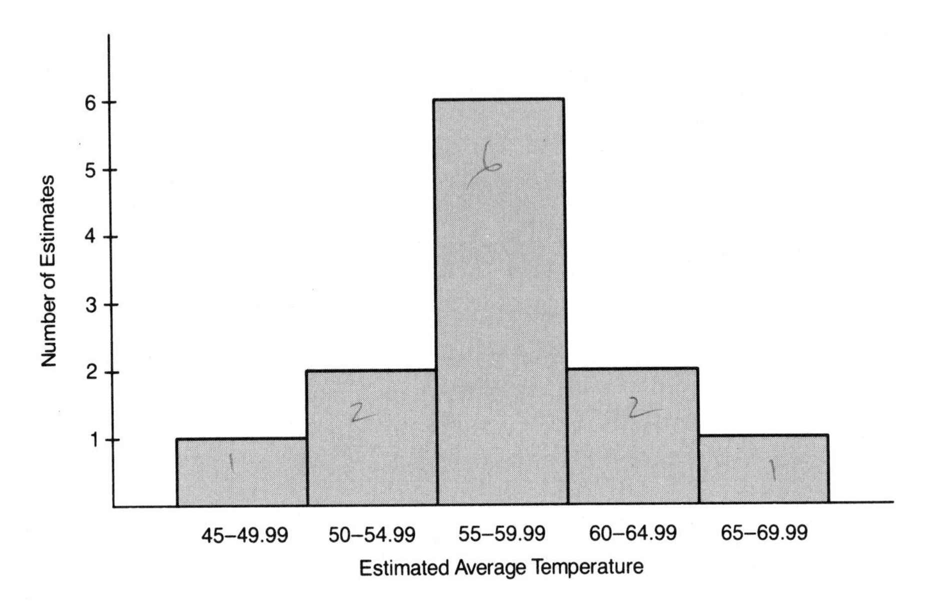

5. Below is a histogram of 1973-1974 total yearly costs for attending 13 randomly selected colleges.

Figure 1-8 Total Annual Costs at 13 Colleges, 1973-1974

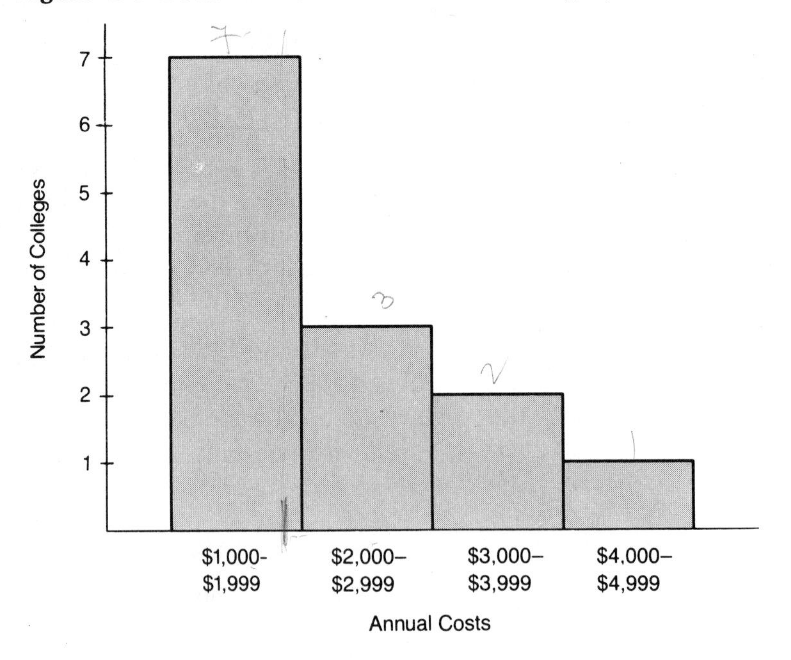

a. Using the balance point idea, estimate the mean.
b. Compute the mean and the median.

6. I once had a statistics class that asked me if everyone had done better than the class average on the previous exam. How would you respond to the class?

7. Below are 50 observations shown as raw data and as a histogram. Which presentation is easier to understand and why?

Raw Data									
1.5	2.5	2.6	3.1	4.7	4.6	3.8	3.5	1.6	3.5
2.4	2.8	2.9	1.7	3.8	3.2	2.5	3.5	4.6	3.7
1.4	2.2	2.3	3.4	4.5	4.7	3.5	3.4	2.8	2.7
1.1	2.5	1.7	3.5	3.4	4.5	1.7	2.5	3.4	2.7
3.5	4.8	4.1	2.8	1.7	1.8	2.6	4.9	1.1	4.1

Figure 1-9 Histogram of 50 Observations

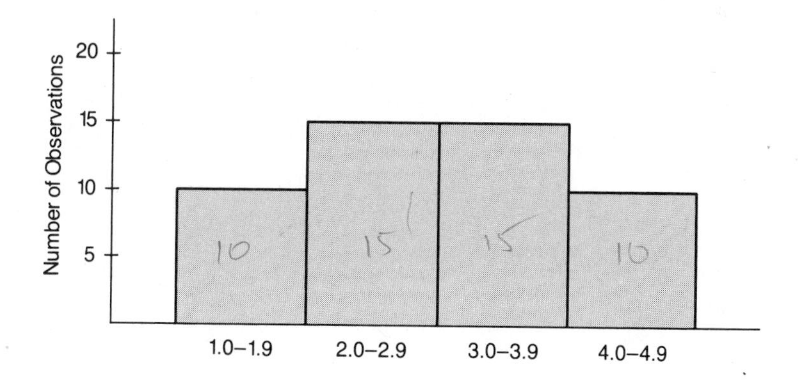

1:4 MEASURES OF SPREAD, DISPERSION, OR VARIABILITY

By the end of this unit you should be able to:

1. Explain the need for a measure of spread.
2. Explain why the range is not an effective measure of spread.
3. Compute the standard deviation for a set of raw data and explain its meaning to others.
4. Estimate which of several histograms exhibit more variance by merely looking at them.
5. Compute the standard deviation for a histogram.

1:4:1 The Range

You are a newly appointed manager in charge of two production lines. Your goals are to increase the average production rates above 50 units per hour and achieve a relatively consistent hourly output. After all, you don't want to produce 100 widgets per hour on Monday and only 10 widgets per hour on Tuesday. This would drive your salespeople up the wall because they could never be sure how long it would take to fill an order. On your first day, your assistant manager reports that the average production rate on the two production lines is 60 units per hour. Have you accomplished your goals? If not, why not? Please write your ideas below.

no - is 60 too high
maybe - 60 is above 50 -
do you have to reduce hours to
meet quota, or do you
need to increase sales?

A mean of 60 units per hour tells you that one of your goals has been accomplished. Can you tell from the mean if you are obtaining a consistent production rate? Perhaps an example will help. Below are hourly production rates taken at random times from the two production lines over a several day period.

Hourly Production Rates for Two Lines

	Department 1	Department 2
	60	80
	59	40
	61	0
	60	120
	60	60
x-bar	60 ✓	60
median	60	60

While both data sets have the same average and median, there is a major difference between the two departments. The hourly production rate in Department 1 is very consistent, whereas the production rate is volatile in Department 2. You can see that the mean or the median can't capture all the information from the raw data. How can we measure the spread or variability in the two data sets?

The simplest measure of spread is the range.

The **range** is equal to the difference between the largest and smallest observations in a data set.

The ranges for the two data sets are 61 − 59 or 2, and 120 − 0 or 120, respectively. This indicates that there is very little spread in the production rate of Department 1 and a large amount of spread in Department 2. It seems to be a useful measure of spread, but it has two very serious weaknesses which reduce its usefulness. Can you determine one of them?

- overproduction/underproduction
- machine/people "burnout"
- lack of work/output on day 3.

only used 2 values

The more obvious weakness is that in computing the range, you use only two values from the data set. The rest of the data is ignored. How can range measure spread in the data if only two observations are used? At best it is a quick estimate of spread. Also, you can easily be misled by the range. Suppose two other data sets are collected and the ranges in both data sets are the same. Can you conclude that the amount of spread within the data is the same? Here are the two data sets.

Data Set A	Data Set B
60	40
60	45
60	50
60	55
40	60
range 20	20

While the range is 20 for both data sets, is the amount of spread the same? I think you'll agree that Data Set B has more spread or variability. After all, *all* the numbers are different. The reason you can be misled by the range is that you only use the smallest and largest values and ignore the other observations.

The second weakness of the range is that it ignores the mean or median in calculating the spread in a data set. A useful measure of spread should tell you how close the observations are to the mean or the median. Earlier we considered a case where the average salary in a company was $100,000. Yet none of the ten observations were close to the mean. If we had computed a measure of spread which used the mean, its value would have been large. This would have warned you that the ten observations are not all close to the mean. You should not have expected to obtain a salary that is very close to the mean salary of $100,000. On the other hand, if the value of spread had been low, then you could have concluded that all the observations are somewhat close to the mean. And if you took a job with the company your salary would be close to $100,000. With the proper measure of spread you can tell how much meaning you can attach to the mean. Is it representative of the data or isn't it?

In summary, a proper measure of spread or variability should use all the data and provide us with information on how close the observations are to the mean or the median (we'll use the former). We turn to such a measure next.

1:4:2 The Variance and the Standard Deviation for Raw Data

The **variance** is the average of the squared differences between *each* observation and the *mean*.

First let's compute the variance in the data for Departments 1 and 2 in the table on page 19 and then discuss some of its properties.

Department 1:

$$\text{variance} = [(60 - 60)^2 + (59 - 60)^2 + (61 - 60)^2 + (60 - 60)^2 + (60 - 60)^2] \div 5$$
$$= 2 \div 5 = .40$$

Department 2:

$$\text{variance} = [(80 - 60)^2 + (40 - 60)^2 + (0 - 60)^2 + (120 - 60)^2 + (60 - 60)^2] \div 5$$
$$= 8{,}000 \div 5 = 1{,}600$$

The variance has the following properties:

1. Unlike the range, all the data are used.
2. Unlike the range, the variance doesn't just measure spread or dispersion, but measures spread or dispersion around the *mean*.
3. The variance can never be negative because we sum the *squared* dispersions around the mean. When the variance equals zero, this means that all the numbers are equal to the mean; there is no dispersion about the mean. So if you ever compute a negative variance you can be sure that you made a math error. You can't have less than no dispersion!

Instead of dividing by the number of observations (five in the above example), it is common practice to divide by the number of observations minus one. One of the goals of statistics is drawing inductive inferences (see Section 1:2:3). Therefore, except in this chapter on descriptive statistics, the variance of a sample is computed because we wish to estimate the variance of the population from which the sample was drawn.

Now let's conduct an imaginary experiment. Suppose you know the variance in incomes in a large neighborhood because you have access to the census data. However, I don't. I would like to know the variance in income level but am unwilling to take a census of the entire neighborhood. So I take a small sample and compute the variance of the sample. After doing this once, I decide to do it a few more times just to be on the safe side. If I compared my sample variances with your known variance for the population from the census data, I would find that most of my sample variances were somewhat smaller than the actual variance for the neighborhood. That would mean that my sample variance tended to *underestimate* the real variance of the neighborhood. Now, we would not want to use an expression for variance that tends to underestimate the actual variance. What we need to do is adjust the expression for the variance of the sample so that it will produce slightly larger variance values. We do this by dividing the numerator of the variance expression by $(n - 1)$, the number of observations minus 1, rather than by n. This will increase the value of the sample variance and bring it closer to the actual variance for the population. So from now on when we compute the variance, the denominator will be $(n - 1)$ rather than n.

You may be wondering why we square the differences between each observation and the mean. Why not just take the differences and sum them? Try a simple example in the space provided below—compute the variance for the data set (3, 4, 5), and see what happens when you omit the squaring operation.

$(3-4)^2 + (4-4)^2 + (5-4)^2 = 2 \div 2 = 1$

$n - 1 = 2$

If you haven't been able to figure it out, this should help.

Observation 1	3
Observation 2	4
Observation 3	5
x-bar	4

Here's how the expression for variance would look if we were to forget to square the differences between each observation and the mean.

$$\text{Variance} = [(3 - 4) + (4 - 4) + (5 - 4)] \div (3 - 1)$$
$$= [(-1) + (0) + (1)] \div 2 = 0$$

As it is obvious that all the values in the above data set are not the same as the mean, the variance cannot be zero. The problem is that the negative values and the positive values cancel each other out. If you use the above expression you will always get a value of zero for the variance.

How can the "positives cancelling the negatives" problem be corrected? One way is to square the differences between each observation and the mean. That's the approach we'll use. Can you think of others?

While the variance is a useful measure of spread in a data set, there is a more meaningful measure. It is derived from the variance (think of it as "son of variance") and is called the standard deviation.

> The **standard deviation** is merely the *positive* square root of the variance. The standard deviation tells you how far observations are away from the mean. *Most* observations will be within one standard deviation of the mean. *Very few* observations will be more than two standard deviations away from the mean.

The terms *most* and *very few* are vague. But with the introduction of Tchebycheff's Theorem in 1:5, you'll be able to replace those terms with precise percentages.

Please compute the standard deviation for the pulse rate data shown below. Remember: (1) first compute the sample mean—x-bar—and (2) the denominator should be $(n - 1)$ not n.

31, 33, 36, 36, 37, 38, 39, 41, 44, 47

The mean is 38.2 and the standard deviation is 4.825. One standard deviation on either side of the mean includes observations between 33.375 (38.2 − 4.825) and 43.025 (38.2 + 4.825). Seven of the ten observations lie in this interval. That's what was meant by "most" of the observations. The interval that includes two standard deviations (38.2 − 9.65 to 38.2 + 9.65) on either side of the mean is 28.55 to 47.85. No observations lie outside this interval. While the number is not always zero, only a few observations will fall outside this interval.

1:4:3 The Standard Deviation for Tabled Data

The variance or standard deviation are both measures of variation around the *mean*. With that idea in mind, look at the following two histograms and determine which histogram will have the greater variance. Each histogram contains ten observations. Write your explanation in the space provided below.

Figure 1-10 Which Histogram Has the Larger Variance?

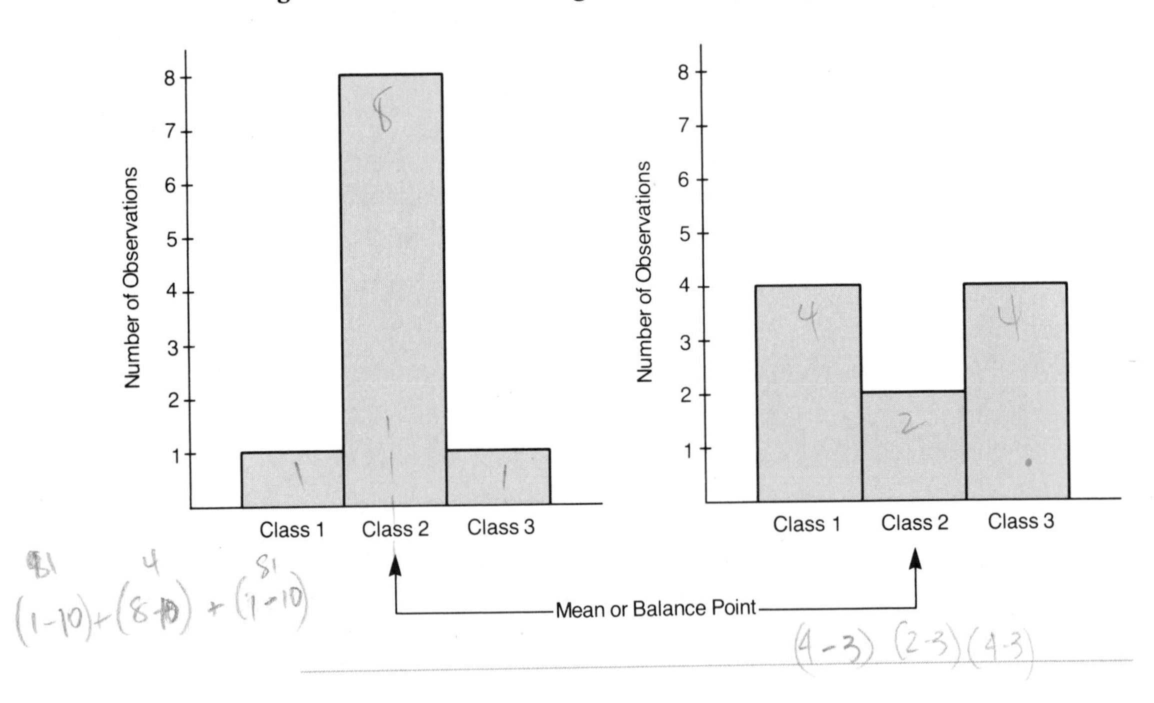

The histogram on the left has less variance. The mean of the histogram is located at the center of Class 2—the balance point. Eight of the ten observations are in the same class as the mean. This tells us that most of the observations are very close to the mean and the variance will be small. By contrast, eight of the ten observations in the other histogram are in Classes 1 and 3, which are as far away from the mean as you can get in a histogram with only three classes. Therefore, the variance will be larger.

On to how to compute the variance and standard deviation for a histogram. Rather than telling you, can you determine the expression yourself with just a little help? Here are the clues:

1. Remember that variance is "almost the average" (the denominator is $n - 1$) of the squared differences between each observation and the mean.
2. Look at what we did when we went from computing the average for raw data to the average for the histogram. Do the same thing but apply those changes to the variance calculation. Both expressions are shown below to guide you.

x-bar for raw data = (Observation 1 + Observation 2 + ... + Observation n) ÷ n

x-bar for Histogram = [(Midpoint of Class 1 × Frequency of Class 1) + (Midpoint of Class 2 × Frequency of Class 2) + (Midpoint of Class 3 × Frequency of Class 3)] ÷ n

Ask yourself, what do the raw data observations correspond to in a histogram and then attempt to develop the proper expression for computing the variance for a histogram. This is difficult, so take your time.

The important point to remember is that variance always measures the squared dispersion around the mean irrespective of whether you are working with raw data or a histogram. With that in mind, did you come up with the following expression?

The variance of a histogram with three classes is

$$[(\text{Midpoint of Class 1} - x\text{-bar})^2 \times (\text{Frequency of Class 1})$$
$$+ (\text{Midpoint of Class 2} - x\text{-bar})^2 \times (\text{Frequency of Class 2})$$
$$+ (\text{Midpoint of Class 3} - x\text{-bar})^2 \times (\text{Frequency of Class 3})] \div (n - 1)$$

Now we can use this expression to compute the variance and ultimately the standard deviation for the pulse rate histogram data. The histogram is reproduced below for your convenience.

Figure 1-11 Pulse Rate Histogram

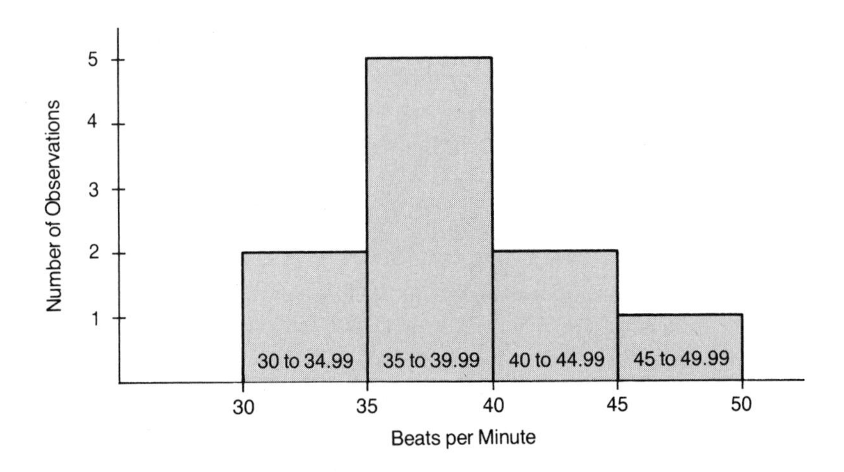

Take a few minutes to calculate the standard deviation in the space provided below.

Here's how you compute the variance for the pulse rate histogram data.

$$\text{Variance} = [(32.5 - 38.5)^2(2) + (37.5 - 38.5)^2(5) + (42.5 - 38.5)^2(2) + (47.5 - 38.5)^2(1)] \div (10 - 1)$$

$$= [(-6)^2(2) + (-1)^2(5) + (4)^2(2) + (9)^2(1)] \div 9$$

$$= (72 + 5 + 32 + 81) \div 9$$

$$= 21.11$$

$$\text{Standard Deviation} = \sqrt{21.11}$$

$$= 4.59 \text{ Beats per Minute}$$

Note that the standard deviation for the histogram data is not exactly equal to the standard deviation for the raw data (4.825 beats per minute). Again, the reason is that we do not use the actual values of the observations in the calculation. We can slightly affect the value of the standard deviation calculated from the histogram by changing the limits of the classes.

Thus, the mean and the standard deviation for the histogram data are 38.5 and 4.59 beats per minute. But what exactly does that tell us? We know that *most* of the observations will be within 4.59 beats per minute of the mean. We also know that very few observations will be more than 9.18 (or two standard deviations) away from the mean. You still probably don't feel comfortable with those vague terms. The next section eliminates the imprecision.

Exercise Set for 1:4

1. Compute the range, variance, and the standard deviation for the following set of data.

   4, 4, 5, 5, 7, 7, 8, 13, 13, 14

2. Which histogram in Figure 1-12 has the greater variance? Why?

Figure 1-12 Two Histograms

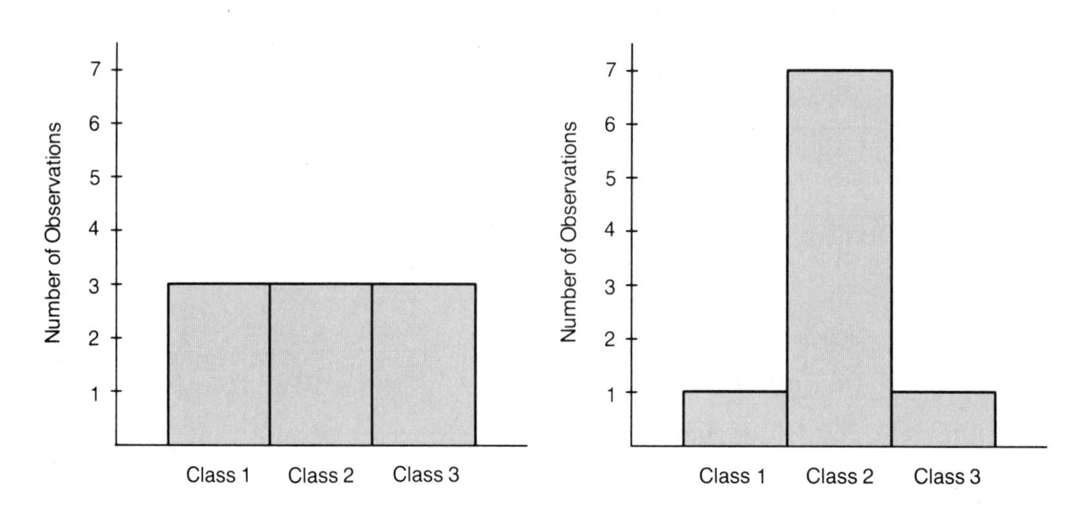

3. Under what conditions, if any, can the variance be less than zero?
4. Compute the variance and the standard deviation for the college costs and temperature of Reno histograms found in Exercise Set 1:3.

1:5 TCHEBYCHEFF'S THEOREM: A FEW NUMBERS IN PLACE OF MANY

By the end of this unit you should be able to:

1. Explain to others why the theorem is useful in reducing a large data set to a few meaningful numbers.
2. Summarize a data set using Tchebycheff's Theorem.

Tchebycheff's Theorem (also known as Chubby Checker's or Chevy Chase's Theorem, depending on your age) is a very powerful tool. It allows you to reduce a large data set to two numbers—the average and the standard deviation—without losing much information. The theorem tells us what percentage of the data will lie at various distances from the mean. We will be able to replace vague words such as "most" and "very few" with precise percentages. In words, the theorem allows us to say that at least a certain percentage of the data must fall within a given number of standard deviations from the mean. More precisely Tchebycheff's Theorem tells us the following:

At least $100 - 100/h^2$ percent of the observations must lie within h standard deviations of the mean.

This expression is true for all values of "h" greater than one. Here's what the expression is telling us.

1. At *least* 55.5 percent of the observations must lie within plus or minus 1.50 standard deviations of the mean (here we substitute a value of 1.50 for h).
2. At *least* 75 percent of the observations must lie within plus or minus 2.00 standard deviations of the mean (here we substitute a value of 2.00 for h).
3. At *least* 88.8 percent of the observations must lie within plus or minus 3.00 standard deviations of the mean.

What this theorem does is provide precise meaning to the terms "most" or "very few." Let's apply the theorem to the pulse rate raw data and see how it works. The mean and standard deviation for the raw data were 38.2 and 4.825 beats per minute respectively.

Number of Standard Deviations	What Theorem States	Actual Percent from Raw Data
1.50	At least 55.5 percent of the data are within 1.50 standard deviations of the mean	$\frac{9}{10}$ = 90 percent
2.00	At least 75 percent of the data are within two standard deviations of the mean	$\frac{10}{10}$ = 100 percent

Practically speaking, there is no set of data that will violate the theorem. The "at leasts" will always be true. Let's try it for a data set that exhibits a large amount of spread. Here's the data.

0, 0, 10, 10, 20, 100, 150, 200, 300, 1,210

$$x\text{-bar} = 200$$

$$\text{Variance} = [(0 - 200)^2 + (0 - 200)^2 + \ldots + (300 - 200)^2 + (1{,}210 - 200)^2] \div 9$$

$$= 136{,}352.95$$

$$\text{Standard Deviation} = 369.26$$

From the theorem we know that at least 75 percent of the observations must fall within two standard deviations of the mean. For this data set two standard deviations around the mean stretch from −538.5 to 938.5. That interval includes 9/10 or 90 percent of the observations. We also know that at least 88.8 percent of the observations must fall within three standard deviations of the mean. This includes an interval from −907.78 to 1,307.7. That interval includes all the observations, which is certainly greater than the 88.8 percent figure. No matter what the mean and standard deviation are, Tchebycheff's Theorem won't be violated.

After doing the exercises below, you should review all the objectives at the beginning of each section. If you haven't mastered them yet, review the material again. Descriptive statistics lays the foundation for formal statistical reasoning. Without mastery you will be lost.

Exercise Set for 1:5

1. Use Tchebycheff's Theorem and the mean and standard deviation for the two histograms in Exercise Set 1:3 to summarize the college cost and temperature of Reno data. Compute the interval—the mean plus or minus two standard deviations—for each data set. At least what percent of the data must lie in this interval? Note: I arbitrarily chose two standard deviations around the mean; you could choose any number of standard deviations (as long as it is greater than one) around the mean.
2. Verify the theorem by computing the mean and standard deviation for the following data set. Show that the "at least" statements of the theorem are true.

   4, 4, 5, 5, 7, 7, 8, 13, 13, 14

2. Basic Probability

2:1 INTRODUCTION TO PROBABILITY

People talk about chances everyday. What are the chances that a city will approve a major redevelopment effort? What are the chances that you will understand this chapter? The official name for chance is **probability**.

If you want to understand probability you must start with the concept of *randomness*. You often hear that quiz show contestants are chosen at random from the studio audience. Pollsters select their subjects at random from some specific group of people. A medical doctor selects patients at random for a medical experiment. These are all examples of random sampling, but randomness goes beyond sampling. If you flip a fair coin twice, you can get either zero, one, or two heads, but you can't predict how many heads you will get the next time you flip two coins. Why? Because the results are random.

> **Random** means that the exact outcome is not predictable in advance. However, a predictable long-run pattern of the outcomes will emerge after many trials (flips of the coin or selection of contestants).

Let's examine both statements. Suppose there are 100 potential contestants sitting in the audience. Eighty are from California and twenty are not. We randomly select ten people to "come on down" and be contestants. Can we predict the exact makeup of the ten contestants? Can we say that eight of the contestants will come from California? Why or why not? Please write your answer on the following page.

I hope you concluded that you could not predict the exact makeup. The reason is that there is randomness. We can't be sure how many of the ten contestants will be from California. When you take a random sample you must account for **margin of error**. However, if there were 100 potential contestants per day (80 from California) and we took ten people at random each and every day, a pattern would emerge over time. Most of the groups of ten would have between six and ten Californians. A few groups might have as few as one or two, although the chances of that happening are small.

If you flip a fair coin twice you cannot make statements such as "this time I'm sure that I'll get two heads." However, if you flip a coin twice many times, a long-run pattern will emerge. Frequently you will obtain one head in two coin flips and less often you will obtain zero or two heads. How often is "frequently" and how do you compute it? That's the next topic.

2:2 BASIC PROBABILITY CONCEPTS

By the end of this unit you should be able to:

1. Explain to others the difference between the relative frequency viewpoint of probability and subjective probability.
2. Compute relative frequency probabilities.
3. Explain the dangers of assessing subjective probabilities.

4. Explain to others the meaning of expected value.
5. Compute the expected value and standard deviation.

Here's a working definition of relative frequency probability or chance:

> The **chance** or **probability** of an event is the percentage of the time the event is expected to happen, provided we select our sample or repeat the experiment under the same conditions many, many times.

Probabilities or chances don't tell us much about a particular toss of a coin or a particular selection of contestants on a single day. But they do tell us what the chances are for different outcomes over the long run. Chances represent the percentage of time a certain event occurs.

2:2:1 Computing Relative Frequency Probabilities

Here's the **basic probability expression**:

> The probability of an event is equal to the number of outcomes for which you wish to compute a probability divided by the total number of possible outcomes.

Here are a few simple examples.

1. There is an urn with 70 red marbles and 30 blue marbles sitting on a table. If you select one marble at random from the urn, what are the chances of getting a red marble?

The total number of possible outcomes or marbles you could obtain on a single draw is 100. But how many of the 100 possible outcomes are you interested in? Only those that are red marbles, and there are only 70 red marbles in the urn. The numerator of the basic probability expression is 70. Thus, the probability of obtaining a red marble is

$$\frac{70}{100} = .70 \text{ or } 70 \text{ percent}$$

The chance of getting a red marble is .70. This means that 70 percent of the time you would get a red marble.

In order to use the basic probability expression, you must be sure that on a single trial (selecting a marble, flipping a coin, or selecting a contestant) *all of the outcomes are equally likely*. Do all the marbles in the urn have an equally likely chance of being selected? Yes. What are the possible outcomes from a single flip of a fair coin? Heads or tails.

Are they equally likely? Yes. If you have a *fair* coin, the chances of obtaining heads or tails are the same—50 percent or .50.

Let's try a somewhat harder example.

2. You flip a *fair* coin twice. What are the chances of obtaining two heads?

Since we are using a fair coin, you can use the basic probability expression to calculate meaningful probabilities. How many different outcomes are possible over two coin flips? Below is a **tree diagram** which you can use to determine the total number of possible outcomes on each and every trial.

Figure 2-1 Tree Diagram for the Fair Coin Example

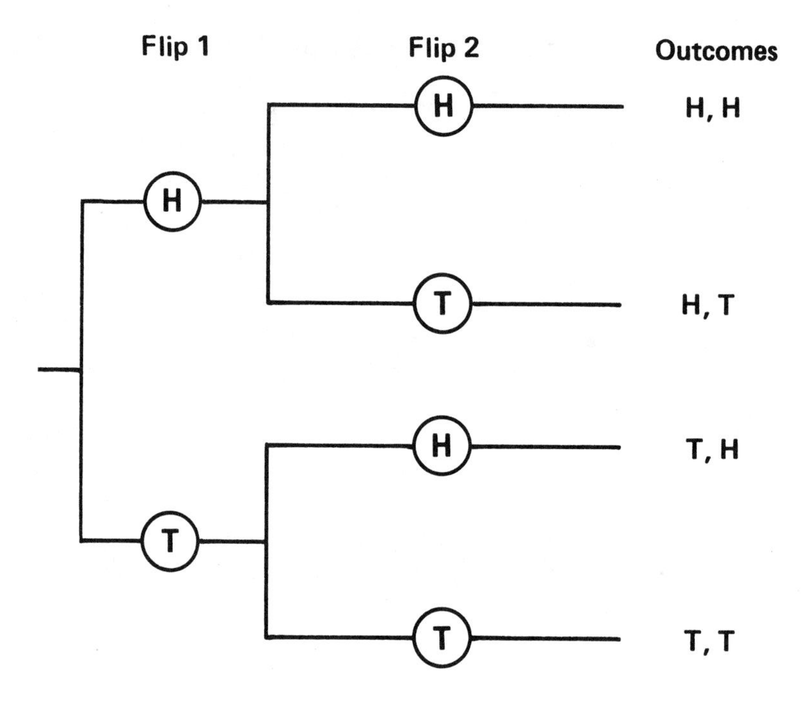

As you can see, there are four possible outcomes when you flip a coin twice (HH, HT, TH, TT). Of the four outcomes, you are interested in those that are two heads (HH). So the numerator is one and the probability of obtaining two heads in two coin flips is $1 \div 4$, or $^1/_4$, or 25 percent. If we repeatedly flipped a fair coin twice, 25 percent of the time we would obtain two heads.

3. What is the probability of rolling a *fair* pair of dice and having the values sum to three?

Figure 2-2 Tree Diagram for Dice Example

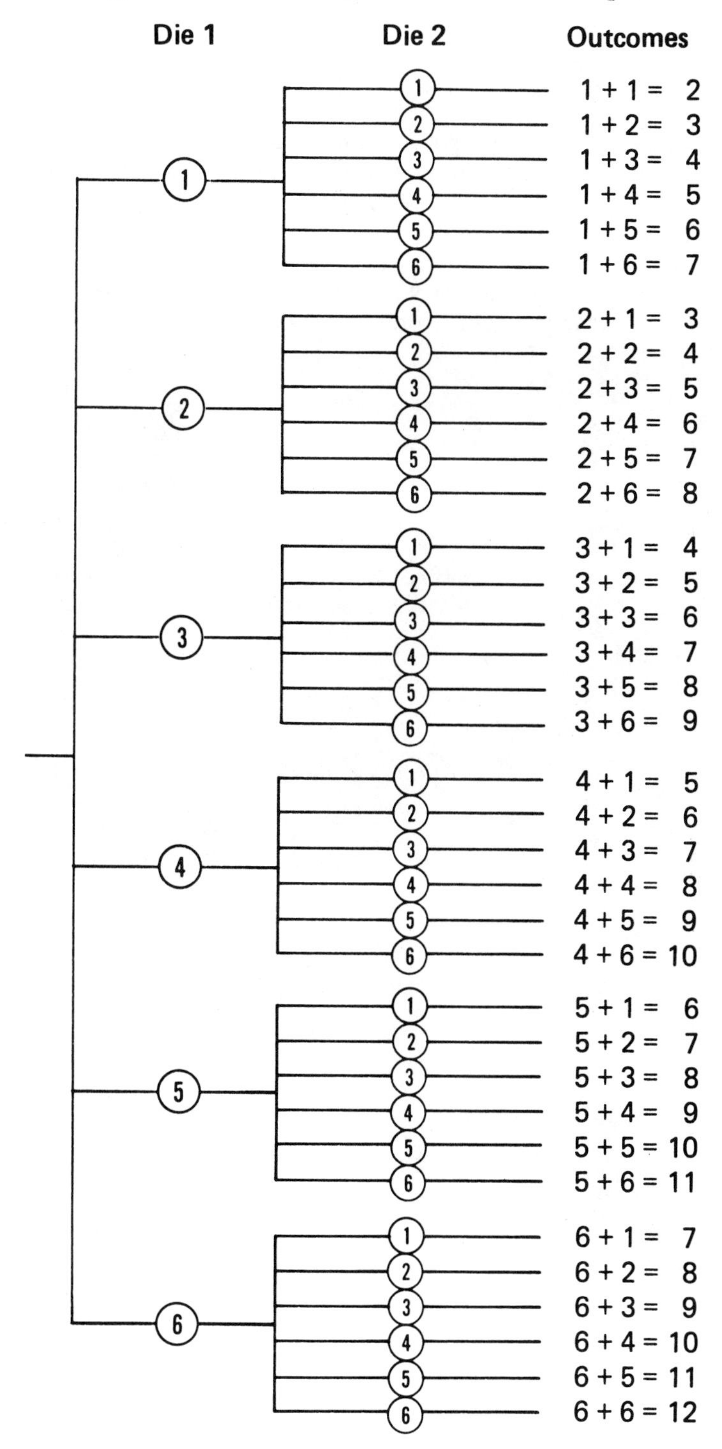

There are 36 possible outcomes. Of those 36 outcomes, only two ($2 + 1 = 3$, $1 + 2 = 3$) sum to three. The probability of obtaining a total of three in rolling a pair of dice is $2 \div 36$, that is, $^{2}/_{36}$, or 5.55 percent of the time.

4. Suppose you have a stock portfolio consisting of two stocks. On any given day each stock may go up, down, or stay the same. What are the chances of both stocks increasing in value today?

In the space provided below, please draw a tree diagram and solve the problem. Don't look ahead.

Figure 2-3 Tree Diagram for Portfolio of Two Stocks

Stock 1	Stock 2	Outcomes
Up	Up	Up, Up
	Stay the Same	Up, Stay the Same
	Down	Up, Down
Stay the Same	Up	Stay the Same, Up
	Stay the Same	Stay the Same, Stay the Same
	Down	Stay the Same, Down
Down	Up	Down, Up
	Stay the Same	Down, Stay the Same
	Down	Down, Down

Out of nine combinations only one has both stocks going up. However, the probability of both stocks going up may not be $^1/_9$. Why? Think about it and write your answer below. What assumption must be true for the basic probability expression to provide meaningful probabilities? This is difficult, so take your time.

Did you remember that the outcomes (movement of each stock each day) must all be equally likely to use the basic probability expression? If either stock does not have an equally likely chance of going up, staying the same, or going down, you cannot use the basic probability expression and the probability is not $^1/_9$. If the underlying assumption isn't true, then the probabilities are worthless.

2:2:2 Rules of Probability

Under what conditions can the probability of an event be less than zero or greater than one? There aren't any.

> Probabilities are numbers between zero (or 0 percent) and one (or 100 percent).

If there is absolutely no chance of an event occurring, it happens zero percent of the time. If it always occurs, it happens 100 percent of the time. Since an event cannot occur less than zero percent of the time, it is impossible to have a negative probability. Since an event cannot occur more than all the time, it is impossible to have a probability greater than 100 percent.

Let's return to the coin flip experiment in Example 2 on page 34. We've already determined that the probability of getting two heads is .25. What are the other possible outcomes of a two coin flip experiment? We could have obtained no heads or one head in two flips. Together with two heads, these account for all the possible events or happenings. Use the tree diagram again and determine the other two probabilities. Insert all three probabilities below.

What do the three probabilities sum to? If you did it correctly, they should sum to one. This leads us to the second rule about relative frequency probabilities or chances.

> The sum of the probabilities of all the possible events that can occur must be 1.0.

2:2:3 Subjective vs. Relative Frequency Probabilities

Up to now we have been computing relative frequency probabilities—probabilities that represent the percentage of time an event will occur.

Probability was first applied to gambling. The relative frequency viewpoint of probability is appropriate for playing blackjack, selecting individuals for a poll, or flipping coins. We can use the basic probability expression to compute relative frequency probabilities provided, of course, that the outcomes on a single trial are all equally likely. This restriction will be loosened up in the next chapter since it is an unrealistic requirement. For instance, is the birth of a girl or a boy equally likely? No, so I'll introduce some other ways to compute relative frequency probabilities in Chapter 3.

While the frequency concept of probability is useful, there are times when it just doesn't apply. This happens when you are asked to assess the chances of a unique, never-before-happened event. What are the chances that your new boss will change the dress policy where you are working? What are the chances that a new microcomputer will outsell its major competitors in the next quarter? What are the chances that your company will acquire another firm this year? Even when there is historical data to guide you, you may choose to ignore it. You may have had many new bosses before and you may have historical data on how many times the dress code was changed, but you may not feel comfortable using that data to assess the chances of this boss changing the dress code. This is **personal** or **subjective probability**.

Personal probabilities are still subject to the two rules of relative frequency probability. Personal probability is a number between zero and one and the sum of all the personal probabilities for all the possible events must be one. You would assign a subjective probability of zero if you were *absolutely* sure that the event *would not* occur; you would assign a personal probability of one if you were *absolutely* sure that the event *would* occur. The more sure you are that the event will occur, the greater should be your probability estimate.

There's good news and bad news regarding personal or subjective probabilities. The good news is that they are very easy to assign. All you do is assign a number that reflects your personal belief about the chances of the event occurring. The bad news is that it may be too easy to assign personal probabilities. It is so easy that often we assign a subjective probability without thinking about why we chose the number we did. Think about these examples:

1. Is the letter *k* more likely to be the first or third letter in an English word?
2. Are you more likely to die from an accidental fall or an accidental discharge of a firearm?

According to Daniel Tversky and Amos Kahneman in a 1973 article in the journal, *Cognitive Psychology*, three out of four people believe that the letter *k* is more likely to be the first letter in an English word. Why? Most people can recall words that begin with the letter *k*. They have difficulty recalling words that have *k* as their third letter. Actually, the letter *k* is two times as likely to appear as the third letter.

How did you answer the second question? Did you think that the chances of accidental death due to firearms was more likely? If you did, you are not alone. Tversky and Kahneman point out that the frequency of dramatic, sensational causes of death which get heavy media coverage is overestimated by most people. Data from the U.S. Public Health Service indicate that the chances of accidental death from falling are many times more likely than from the accidental discharge of a firearm.

Subjective probabilities are subject to all the biases that we humans have. If we happen to be in an optimistic mood, we may provide probability estimates that are overly optimistic. Sometimes the probability estimates are influenced by our ability to recall whether similar events have occurred recently. We would tend to overestimate the chance of being killed by a drunken driver if we had recently read an article on drunken drivers. We are simply not good probability estimators. Pogo, the comic strip character, said it best: "We have met the enemy and we is they."

2:2:4 Expected Value and Standard Deviation

Many states have instituted lotteries as a way of raising revenues. Other groups such as The Publisher's Clearinghouse use lotteries to get us to subscribe to magazines. Suppose you purchase a $5 ticket to play a lottery. Here are your chances of winning the various prizes.

Amount of Prize	**Number of Winners per Million Entries**	**Probabilities**
$500,000	1	.000001
$100,000	5	.000005
$10,000	10	.00001
$1,000	1,000	.001
nothing	998,984	.998984

From the relative frequency viewpoint of probability, you know that one one-millionth of the time you will win the $500,000 prize. Most of the time you will win nothing. If you were to play this lottery many times, what would your *average* winnings be? In the *long run*, what can you *expect* to win each time you play the lottery?

The **expected value** is a measure of central tendency for probability data. It is a weighted average. You multiply each outcome by its probability and sum them.

Here is how you compute the expected winnings for the lottery.

Expected winnings = ($500,000 × .000001) + ($100,000 × .000005)
+ ($10,000 × .00001) + ($1,000 × .001)
+ ($0 × .998984)
= $2.10 per lottery ticket

Expected value is the long-run average; it is your average winnings per lottery if you played the above lottery many times. If you have to pay $5 for the ticket but the average winnings per ticket is $2.10, you can see how lotteries are revenue generators for states.

If you play the lottery many times, your average winnings will be $2.10 per lottery. How meaningful is expected value if you play the lottery just once? It depends. How meaningful is the average for raw data? What other measure do you need before you could decide how meaningful the average was? Please write your answer below.

Did you remember that you needed to compute the variance or the standard deviation? You must do the same thing here. You would need to compute the *standard deviation of the probability data.* Now, if the standard deviation is small in comparison to the expected value, what could you conclude?

You should conclude that the expected value is not only meaningful in the long run, but it is meaningful for a one-time purchase of a lottery ticket.

Here is how you compute the variance. Whether you have raw, histogram, or probability data, variance is still the average squared dispersion of the data around the mean. When working with probability data, you must compute the weighted average by multiplying each of the squared differences *by their respective probabilities.*

$$\begin{aligned}\text{variance} = {} & (500{,}000 - 2.10)^2(.000001) \\ & + (100{,}000 - 2.10)^2(.000005) \\ & + (10{,}000 - 2.10)^2(.000010) \\ & + (1{,}000 - 2.10)^2(.001) \\ & + (0 - 2.10)^2(.998984) \\ = {} & 301{,}995.59\end{aligned}$$

$$\text{standard deviation} = \$549.54$$

As you can see, the standard deviation is very large as compared with the $2.10 expected value. The expected value is not a meaningful figure to use if you play this lottery only once. Your actual winnings from buying only one lottery ticket may vary significantly from the $2.10 expected value figure. As you can see from the lottery probabilities, most of the time you will win nothing—not even $2.10. Once in a very long while you will win a prize. Nevertheless, the expected value of $2.10 is still a meaningful number if you play the lottery many times.

You need to know the standard deviation before you attach too much meaning to the expected value on a one-shot deal. You first learned that lesson when you applied Tchebycheff's Theorem to raw and histogram data. Now you know the rest of the story; it applies to probability data as well.

Insurance companies also find expected value useful. An insurance company offers a 42-year-old man a one-year $100,000 term policy for $222. Suppose the actuarial tables tell us that the chance that a 42-year-old man will die within the next year is two in a thousand. In the long run (over many thousand policyholders), what is the expected revenue to the company?

There are only two possible outcomes—the policyholder lives or dies.

Possible Outcomes	Revenue Flows to Insurance Company	Probability
1. policyholder lives	+$222	$\frac{998}{1{,}000} = .998$
2. policyholder dies	+$222–$100,000 = −$99,778	$\frac{2}{1{,}000} = .002$

$$\text{expected value} = (+\$222 \times .998) + (-\$99{,}778 \times .002)$$

$$= \$22 \text{ per policyholder}$$

$$\text{variance} = (222 - 22)^2(.998) + (-99{,}778 - 22)^2(.002)$$

$$= 19{,}960{,}000$$

$$\text{standard deviation} = \$4{,}467.66$$

Obviously the insurance company does not make $22 on every policyholder; you can tell this from the large standard deviation. Most of the time the insurance company is ahead the $222 premium. Occasionally the insurance company will have to pay out the face value of the contract. Over a large number of policyholders, the insurance company will average a $22 contribution to profit and overhead. After all, insurance companies are in business to make a profit.

Exercise Set for 2:2

Problem set for basic probability.

1. One card is drawn at random from a regular deck of playing cards. What is the probability of drawing a ten?
2. An urn contains five marbles. Three marbles are red and two marbles are blue. If you select two marbles *without replacing* the first before drawing the second, what are the chances of obtaining two red marbles? Draw a tree diagram to solve the problem.
3. If you replaced the marbles after each draw would the probability change? Should it get larger or smaller and why? What's the new probability?
4. If the birth of a boy or a girl is equally likely, what are the chances that all four children in a family will be boys? Draw a tree diagram to solve the problem.

Problem set for expected value and standard deviation.

1. In families with four children, it is possible to have from no boys to four boys. You have already computed the probability of having four boys in Problem 4 above. Now compute the probabilities of having zero, one, two, and three boys. What must the five probabilities sum to and why? Compute the expected number of boys in families having four children. Compute the standard deviation for the number of boys in families having four children.
2. The accounting department in a firm has estimated the following profit figures for sales of a new supercomputer.

Number of Supercomputers Sold per Quarter	Profits for the Quarter
1	$25,000
2	45,000
3	60,000
4	75,000

The sales department has provided you with the following subjective probabilities of selling one to four machines per quarter.

Number of Supercomputers Sold	Probability
1	.1
2	.5
3	.3
4	.1

What is the expected profit to the firm in the upcoming quarter? Does that mean that they will actually get that profit in the upcoming quarter? Discuss.

3. If the standard deviation in the profit for the upcoming quarter were negative, what could you conclude? Compute the standard deviation in the profit for the upcoming quarter.

2:3 INTRODUCTION TO STATISTICAL INDEPENDENCE

By the end of this unit you should be able to:

1. Explain to others, in your own words, what statistical independence is.
2. Distinguish, using only logic, between situations that are statistically independent and dependent.
3. Explain to others why, in many real-world situations, logic cannot be used to determine if events are statistically independent or not.
4. Distinguish between two types of probabilities—conditional and unconditional or simple probabilities.

One of the major goals in analyzing data is *looking for relationships*. If you take megadoses of vitamin C, will you reduce the probability of catching colds? Are families with higher incomes more likely to support an all-weather sports stadium? Looking for relationships begins with the concept of statistical independence.

Two events are **statistically independent** if the occurrence of one event does not affect or is not affected by the occurrence of the other event.

2:3:1 When Logic Works

Suppose that 80 percent of all students who have taken statistics (also known as sadistics) have passed the course. Randomly select a student from a statistics class. Given no other information, what are the chances that he or she will pass the course? The answer is 80 percent. Here are two additional pieces of information.

First Piece of Information: The student is over six feet tall.
Second Piece of Information: The student spends over 30 hours a week studying statistics.

Which piece of information, if any, should cause you to change your 80 percent probability figure that the student will pass? Please write down your answer with a brief defense below.

You probably said that you would not want to change the 80 percent figure once you learned that the student was over six feet tall. After all, a student's chances of passing are not affected by his or her height. On the other hand, if a student studies over 30 hours per week, it should increase his or her chance of passing the course.

You have just concluded that the student passing the course and the student being over six feet tall are statistically independent events. However, the student passing the course and the student studying over 30 hours per week are statistically dependent events.

The key is whether *upon the receipt of the second piece of information you felt that you should revise the initial 80 percent figure.* When you believe it is necessary to revise the probability, you are arguing that the two events are statistically dependent. The second piece of information therefore is valuable and should be used to increase or decrease the initial probability figure. When you believe that the information is irrelevant, you should not revise your initial probability figure. You have just concluded that the two events are statistically independent. Irrelevant information should *not* cause you to change your probability assessment.

2:3:2 Conditional Probabilities

Next we need to distinguish between two types of probabilities—unconditional or simple probabilities and conditional probabilities.

Here are the three probability statements from the statistics class example presented earlier:

1. Select a student at random from the class. With no other additional information, what are his or her chances of passing the course?
2. Select a student at random from the class. Before you assess his or her chances of passing, you are told that the student has studied more than 30 hours per week in the class.
3. Select a student at random from the class. Before you assess this student's chances of passing, you are told that he or she is over six feet tall.

All three statements deal with the probability of a single event happening—passing the course. What is different about the three statements?

In the last two statements you are given additional information about the student. In the first statement you are given none.

The first statement is a **simple** or **unconditional probability**. This is a probability of a single event happening. All the probabilities we have calculated up to this point in this chapter have been simple or unconditional probabilities.

The last two statements are called **conditional probabilities**. Similar to the first probability, they are probabilities of a single event (passing the course). However, now we're looking for the probability of a single event (call it Event A) *conditional* upon receiving a second piece of information (call it Event B). Event B in Statement 2 is that the student studies more than 30 hours per week. Event B in Statement 3 is that the student is over six feet tall.

Now, let's use only logic to assign probability values to the three statements. Based upon historical data, we know that 80 percent of all students pass the course. Therefore, with no other additional information, the probability of passing the course is 80 percent.

The conditional probability for Statement 2, the probability of passing the course given that the student studies more than 30 hours per week, should be over 80 percent. Exactly how much to increase the probability is not clear at this point. But we can argue logically that it should be increased because students who study more than 30 hours per week do better than students in general. In the next section we'll determine how to revise the probability figure.

The third statement is also a conditional probability because you have been given a second piece of information about the student (over six feet tall). However, this information is irrelevant and you wouldn't revise your 80 percent probability estimate. Thus, the conditional probability for Statement 3, the probability of passing the course given that the student is over six feet tall, is still 80 percent.

These ideas lead to the following definitions:

> Events A and B are **statistically independent** if there is no change in the original probability of Event A happening upon receipt of the information found in Event B.

> Events A and B are **statistically dependent** if there is a change in the original probability of Event A happening upon receipt of the information found in Event B.

This is merely a restatement of the initial definitions of statistical independence and dependence.

When the conditional probability of Event A given Event B is equal to the simple probability of Event A, Event B must be irrelevant since we don't revise the conditional probability. Thus, Events A and B are statistically independent of one another. Events that are statistically independent are not related to one another.

When the conditional and simple probabilities are *not* the same, it means that the additional information (Event B) caused you to revise the probability of Event A happening. Thus, Events A and B must be statistically dependent. Events that are statistically dependent are related to one another.

Except for trivial problems, we must go beyond logic to determine if two events are statistically independent or not.

2:3:3 When Logic Fails

We can't always use logic to determine independence because real-world problems are more complicated than our simple example. Is a person's chances of being promoted within an organization related to his or her sex? If it is, the company may be in violation of the 1964 Civil Rights Act. Here, logic won't work. We will need to collect data and test for statistical independence.

Are the chances of getting a cold affected by vitamin C intake? Logic is not enough, since you can find strong arguments to support and reject Dr. Linus Pauling's theory that vitamin C reduces the chances or severity of colds. Again, we will have to go beyond logic and collect data to determine if the two events are dependent or not. That's our next topic.

2:4 STATISTICAL INDEPENDENCE AND CONTINGENCY TABLES

By the end of this unit, you should be able to:

1. Construct contingency tables from word problems.
2. Determine, using a quick test, if two events *appear* to be related or not.
3. Explain what's wrong with the test (why it can be misleading).

A national women's group has decided to sue a large organization for discrimination in its promotional practices. The group believes that women have been systematically denied promotion. Their lawyers have obtained access to the organization's personnel records for the past five years. However, the courts have limited their access since the information is of a highly confidential nature. The courts will allow the group to select a random sample of 200 employees hired five years ago and to follow their organizational careers. The evidence from the sample will be used to determine if the organization is practicing discrimination. Both sides have agreed to the procedure.

Here's the data for the sample of 200 randomly selected employees.

200 Employees			
120 Males		80 Females	
50 Promoted	70 Not Promoted	15 Promoted	65 Not Promoted

A contingency table is an effective way to organize two-dimensional data. The two dimensions in the present study are:

1. the sex of the employee
2. the promotional status of the employee

Each dimension must be divided into mutually exclusive and exhaustive categories. That means that you must account for all 200 employees. The sex of the employee can only have two categories, male and female. You could subdivide promotional status into more than two categories; for example, never promoted, promoted once, promoted two or more times. I've only used two categories here—promoted and not promoted. These categories are exclusive and exhaustive.

The contingency table is shown below. I've already inserted the total number of males (120) and females (80) as well as the total number of promoted people (65) and nonpromoted people (135).

Contingency Table for Sex Discrimination Suit

	Male	**Female**	**Total Employees**
Promoted			65
Not Promoted			135
Total	120	80	200

This is a two by two contingency table. Each dimension of the contingency table has two subcategories which account for four cells. What numbers should be placed in each of the cells? Please fill them in without looking ahead.

The upper-left cell should contain the number of males who were promoted—50 individuals. Likewise, the upper-right cell should contain the number of promoted females—15 individuals. You can determine the remaining cell entries. The completed table looks like this.

Contingency Table for Sex Discrimination Suit

	Male	Female	Total
Promoted	50	15	65
Not Promoted	70	65	135
Total	120	80	200

Are the two events (sex and promotional status) statistically independent? We know that events are independent when we do not revise our probability of Event A happening upon the receipt of additional information, Event B. Let's use that idea as a quick test.

What's the probability of selecting a male at random from the sample? Using the basic probability expression, the probability of selecting a male employee is

$$\frac{120}{200} = .60$$

Now suppose I tell you that the person selected has been promoted. If the probability that the person is a male changes from the original value of .60, the two events are related. If the probability doesn't change, then the two events are not related. Since we can't use logic, let's *compute* the conditional probability of selecting a male employee given that the individual has been promoted.

You will not need any new formulas to compute a conditional probability. Continue to use the basic probability expression because, on a single trial, all the employees have an equally likely chance of being selected. The denominator refers to the total number of possible events. When we computed the probability of selecting a male employee, the denominator was 200 people. When we compute the conditional probability, the denominator is *revised downward* from all 200 people to only those that are promoted, namely, 65 individuals. Once we know that the individual has been promoted, we focus on only the 65 people that have been promoted.

The numerator is still the number of outcomes for which you wish to compute the probability. Of the 65 individuals who have been promoted, how many are we interested in? Only those that are male—50 individuals. So the conditional probability that the selected employee is male, given that the individual has been promoted, is

$$\frac{50}{65} = .769$$

To summarize, the simple probability of a randomly selected employee being male was 60 percent while the conditional probability of the selected employee being male was 76.9 percent given that the individual had been promoted. The two events *appear* to be statistically dependent.

Let's compute another set of simple and conditional probabilities. In the space below, please compute the simple probability of selecting a female employee and the conditional probability of selecting a female employee given that we selected a promoted individual.

The probability of selecting a female employee is

$$\frac{80}{200} = .40$$

The probability of selecting a female employee given that we selected a promoted individual is

$$\frac{15}{65} = .23$$

When you know that the individual who has been selected has been promoted, the probability of that person being a female drops from 40 percent to 23 percent. The two events *appear* to be statistically related.

Who's being discriminated against? Is it men or women? While you probably already know the answer from the previous calculations, let's compute two other conditional probabilities that will tell the story directly. What are the probabilities of being promoted given that you are a male and given that you are a female?

1. The probability of being promoted given that you are a male is

$$\frac{50}{120} = .417$$

2. The probability of being promoted given that you are a female is

$$\frac{15}{80} = .188$$

The two conditional probabilities are not the same. Neither probability is equal to the simple probability of being promoted—$^{65}/_{200}$ or .325. This tells us that without knowing a person's sex, the chance that a randomly selected person was promoted is 32.5 percent. However, once we find out that the person is a male, the probability increases to 41.7 percent. It drops for a female to 18.8 percent. It *appears* that the company is discriminating against women.

The organization could make two arguments in its own defense. They might argue that while it is true that men seem to have a greater chance of being promoted than women, there are extenuating circumstances. They might argue that the men had higher grades in school and thus were better qualified. The organization might look for other reasons beyond systematic discrimination that might account for the differences in the percentages of each sex promoted. A judge would rule on the validity of these alternative explanations.

A more powerful argument is that the results are only based upon a sample of all the employees. Imagine that the company had not discriminated in its promotional practices. Thus, if we studied *all* the employee records we would find that the conditional probabilities of being promoted given the employee's sex were the same for males and females. Now we take a small random sample of employees from the company. We again compare sex and promotional status. Must the two conditional probabilities be the same for the sample?

You probably would agree that within a sample you might not expect the two probabilities to be the same. There could be a little difference depending upon the particular individuals selected for the sample. We would expect the conditional probabilities to be close but not necessarily the same. When we take a small sample and attempt to draw conclusions about the population from which the sample was taken, we must allow room for a **margin of error**. This is the basis of inductive inference, which will be discussed in Chapter 4.

Inductive inference is making educated guesses about a population based upon a small sample. Does the Gallup polling organization interview all 100 million voters each time they wish to conduct a presidential poll? No! They only interview about 1,500 voters and then make an inductive inference about all 100 million voters. The Gallup organization knows that the percentage of the 1,500 voters favoring a candidate will not be identical to all 100 million voters. They must and do allow for a margin of error. Nevertheless, they are confident that the two percentages will be close.

The quick test discussed here ignores inductive inference. Thus, you should not rely upon it in making your final judgment. Remember, it's only an approximation.

The vitamin C controversy. A scientist wishes to study the effect of vitamin C on the severity of colds. She randomly selects 600 people and randomly subdivides the group into four groups of 150 each. Each group will receive one of the four treatments shown below.

Treatment 1	no additional supplement
Treatment 2	1,000 mg. of vitamin C as a daily supplement
Treatment 3	5,000 mg. of vitamin C as a daily supplement
Treatment 4	10,000 mg. of vitamin C as a daily supplement

The groups of 150 are randomly assigned to the four treatments. The individuals within a group are not aware of what treatment they are receiving. Even the researcher doesn't know who is receiving what treatment. This is called the "double blind" experiment and is frequently used in medical research. The 600 people have agreed to follow a prescribed meal plan which provides for 100 mg. a day of natural vitamin C. Each subject is examined every week by the research team for colds. At the end of the winter the data are summarized. Here's some hypothetical data.

Vitamin C Study

Treatment Group 1	50 had no colds 50 had one to two colds 50 had more than two colds
Treatment Group 2	60 had no colds 50 had one to two colds 40 had more than two colds
Treatment Group 3	80 had no colds 40 had one to two colds 30 had more than two colds
Treatment Group 4	100 had no colds 30 had one to two colds 20 had more than two colds

Here is the completed three (number of colds) by four (treatment groups) contingency table.

Contingency Table for Vitamin C Study

	Group 1	Group 2	Group 3	Group 4	Total
No Colds	50	60	80	100	290
One to Two Colds	50	50	40	30	170
More than Two Colds	50	40	30	20	140
Total	150	150	150	150	600

In the following space, please determine these probabilities:

1. The probability of having no colds
2. The probability of having no colds given that you are member of Group 1
3. The probability of having no colds if you are member of Group 4

What conclusion can you reach about vitamin C and the number of colds? Do they *appear* to be statistically related or not?

Here are the answers.

1. The simple probability of having no colds is

$$\frac{290}{600} = .483$$

2. The conditional probability of having no colds given that you are a member of Group 1 is

$$\frac{50}{150} = .333$$

3. The conditional probability of having no colds given that you are a member of Group 4 is

$$\frac{100}{150} = .666$$

The quick test *suggests* that the events are related to one another. The chances of having no colds seem to increase as you take more vitamin C.

Exercise Set for 2:4

1. Two hundred people are asked their position on a new domed sports stadium. Their responses have been classified according to their income level. Of the 80 people with gross incomes in excess of $30,000 per year, 50 are in favor of the stadium. Of the remaining 120 lower income level individuals, only 40 are in favor of the stadium.

   Construct a contingency table and determine if income level *appears* to affect position on the stadium.

2. We randomly sample 1,200 voters throughout the United States. Of the voters selected, 600 are found to be Democrats; 400, Republicans; and 200, Independents. Of the 600 Democrats, 500 are not in favor of increasing military spending. Of the 400 Republicans, 250 are in favor of increasing military spending. The 200 independents are split evenly. Construct a contingency table displaying this data.

3. Determine the following from the contingency table from Problem 2:

   a. The probability of selecting a Democrat
   b. The probability that a voter is in favor of increasing military spending given that he or she is a Republican
   c. The probability that a voter is a Democrat given that he or she is in favor of increasing military spending

4. Below is a two by two contingency table with some of the data. Fill in the missing data so that the two events appear to be independent.

	Promoted	**Not Promoted**	**Total**
Male			120
Female			80
Total	60	140	200

You may find this problem more difficult than the others, so here's a hint:

Remember what independence means. It means that when you receive additional information about an individual (his or her sex) you will not change the probability that he or she has been promoted. What is the probability of being promoted? What should the probability of being promoted given you are a male be if the two events are independent? What should the probability of being promoted given you are a female be if the two events are independent? Use this to solve the problem.

2:5 JOINT PROBABILITY

By the end of this unit you should be able to:

1. Distinguish between simple, conditional, and joint probabilities.
2. Extend the use of the basic probability expression for computing joint probabilities.

We have computed two types of probabilities—simple and conditional. Both deal with the probability of a *single event* happening. Joint probabilities deal with *two or more events* happening. For example, in the case of the company that appeared to be discriminating against women, what is the probability of selecting a promoted male? We are interested in the joint probability that the person selected is male *and* promoted.

We compute joint probabilities using the basic probability expression. What is the denominator for this probability? Since it is not a conditional probability, it is all 200 people. Who are we interested in? Only those that are male *and* promoted. Look at the contingency table on page 51. You'll see that there are 50 individuals who are male and promoted. Thus, the joint probability is

$$\frac{50}{200} = .25$$

What is the joint probability that a person randomly selected from the vitamin C study is a Group 1 member and had no colds? Look at the contingency table on page 55. The denominator is all 600 subjects. Of those, how many are Group 1 members with no colds? Fifty. Thus, the joint probability is:

$$\frac{50}{600} = .083$$

It is easy to distinguish between a simple probability and a joint probability. But some people have difficulty differentiating between joint and conditional probabilities. A joint probability is the probability of two or more events happening (Event A *and* Event B). Here's an example—the probability of selecting a promoted male. The individual we select might be male or female and might have been promoted or not. We want to know the probability that the person is male (Event A) *and* promoted (Event B). Joint probabilities can involve more than two events, for example a 35- to 44-year-old Democrat in favor of increased military spending. We have three events—35 to 44 years old *and* a Democrat *and* in favor of increased spending.

A **conditional probability** is the probability of a *single* event happening (Event A) given additional information (Event B has happened). For example, the probability that a randomly selected employee has been promoted given that he is a male; the probability that a voter is in favor of increased military spending given that he or she is a Democrat. These are conditional probabilities—we are interested in the probability of Event A (selecting a promoted person, selecting a person who's in favor of increased spending), knowing that Event B has happened (the person is male or Democrat). We want to know the probability of a single event in each case, but they are not *simple* probabilities because of the additional information.

2:6 REVISING PROBABILITIES: THE IDEA BEHIND BAYES' THEOREM

By the end of this unit you should be able to:

1. Explain why it is dangerous to revise probability estimates on intuition alone.
2. Revise probabilities by using contingency tables.

We can use contingency tables for more than determining statistical independence. They can be used to determine how much we ought to revise probability estimates when we receive additional information. This procedure is based on a formula that statisticians call Bayes' Theorem. But the logic of the procedure is simple and I think you'll understand it better if I explain it without using the formula.

If we have the simple probability that Event A will happen, and we receive the additional information that Event B has happened, we now have a conditional probability. If Events A and B are statistically dependent, the new conditional probability is different from the original, simple probability. The revised probability is more accurate than the original one, since it contains more information.

Let's see how good you are at revising probabilities using intuition. Imagine ten urns, each of which contains 100 poker chips. Four of the urns contain 70 red chips and 30 blue chips. Six contain 70 blue chips and 30 red chips. The chips are randomly distributed within each urn. If you select an urn at random, what are the chances that it contains 70 red chips and 30 blue chips? Since four of the ten urns contain 70 red chips and 30 blue chips, the probability is $^4/_{10}$ or 40 percent.

Now comes the hard part of the problem. Suppose I select ten chips from the urn, with replacement. This means I put each chip back after I note its color, so the urn always contains 100 chips. Out of the ten chips, six are red and four are blue. Based on this sample of ten chips, you can decide to

1. *increase* the probability from the present 40 percent that the urn contains 70 red and 30 blue chips; or,
2. *decrease* the probability from the present 40 percent that the urn contains 70 red and 30 blue chips; or,
3. keep the present 40 percent probability figure.

Now that I have selected six red and four blue chips, what is your probability estimate that the urn originally contained 70 red chips and 30 blue chips? Circle one of the figures below. Then in the space provided, develop a brief argument as to why you selected the probability. *Don't do any math* in developing your estimate.

0% 10% 20% 30% 40% 50% 60% 70% 80% 90% 100%

____

____

____

____

____

Most people believe that they should increase their original 40 percent estimate since six of the ten chips were red. They generally increase the probability from 40 percent to 50 or 60 percent. Is that what you did?

The actual probability is about 80 percent. To get the exact value you need to know more about probability distributions, which we won't discuss until Chapter 3. What is important is that most people tend to *underestimate* the amount of probability revision. They tend to undervalue the new information (Event B) and do not revise the probability enough. Problem solvers call this the **conservatism bias.** Other times we overestimate the amount of probability revision. We can use contingency tables to tell us just how much the additional information is worth and to help us determine what is the correct amount of probability revision. Here's an example of how it works.

How good are seismic tests? Before a well is drilled, oil companies frequently conduct seismic tests to determine if oil is present. They drill a hole in the ground, insert a dynamite charge, and detonate it. Based upon the sonic waves, they get an indication as to whether oil is down there. If the test results indicate an open structure, this means that either oil, water, or air is present. A closed structure tends to indicate no oil. Oil companies know that seismic tests are not perfect. How good are they?

Suppose an exploration team has obtained the following test data from other companies drilling in the same oil field. Twenty percent of all wells drilled have struck oil. Of those that struck oil, 70 percent had obtained an open structure in their seismic test. Of those that had not struck oil, 75 percent had found a closed structure.

The simple probability of striking oil with no test is 20 percent. What are your chances of striking oil given you obtain an open structure from your seismic test? Without doing any math, make an educated guess and circle it below. Jot down a brief defense in the space provided.

0% 10% 20% 30% 40% 50% 60% 70% 80% 90% 100%

First, let's construct a contingency table for this problem. We are dealing with two dimensions—(1) the results of the seismic test and (2) whether oil is present. Each dimension must be subdivided into mutually exclusive and exhaustive subcategories. The seismic findings dimension is subdivided into (1) open structure and (2) closed structure. The presence of oil dimension is subdivided into (1) find oil and (2) do not find oil.

Contingency Table for Oil Drilling

	Find Oil	Do Not Find Oil	Total
Open			
Closed			
Total			

You might think that in order to construct a contingency table we need to know the total number of wells in the oil field. Actually, it makes no difference. You can assume any number and develop a contingency table.

Let's assume a total of 200 wells and construct the contingency table using the following probability data.

1. Twenty percent of all wells drilled struck oil.

2. Of those that struck oil, 70 percent obtained an open structure in their seismic test.

3. Of those that did not strike oil, 75 percent had found a closed structure.

Can you complete the contingency table? Please try it before reading on.

There are a total of 200 wells. We know that 20 percent (.20 × 200) or 40 found oil. This means that there were 160 dry wells. Of the 40 producing wells, you were told that the seismic test had found an open structure for 70 percent (.70 × 40) or 28 wells. Of the 160 dry wells, you were told that 75 percent (.75 × 160) or 120 had a closed structure.

200 Wells			
40 Found Oil		160 Dry	
28 open structure	12 closed structure	40 open structure	120 closed structure

Now we can complete the two by two contingency table.

Contingency Table for Oil Drilling

	Find Oil	Do Not Find Oil	Total
Open	28	40	68
Closed	12	120	132
Total	40	160	200

What is the probability of finding oil given that you obtain an open structure seismic reading? We will use the basic probability expression. Remember, the denominator must be revised downwards from all 200 wells to only those for which we found an open structure—68. Of those 68 wells, we found oil in 28 of them. Thus, the probability is

$$\frac{28}{68} = .41$$

Most people would estimate around 70 percent. However, the contingency table tells us that the probability of finding oil once you find an open structure is only 41 percent. Most people are lousy probability revisers.

If you look at the contingency table you can determine why the probability is so low. Rewrite the conditional probability of finding oil given an open structure as

$$\frac{28}{(28 + 40)}$$

Remember, of those 160 dry wells, 75 percent had found a closed structure. This means that 25 percent of the dry wells had obtained an open structure seismic reading. That's not a very large percentage. However, even 25 percent of 160 dry wells is a large number of wells—40 to be exact. The majority of wells with an open structure are dry. Thus, the probability of finding oil given an open structure is less than most people would guess.

Revising probabilities using contingency tables is based on Bayes' Theorem. The idea behind it is simple. Since we have new information we can eliminate some cells in the contingency table. Then we revise the numerator and denominator of the basic probability expression using only the remaining cells. Since we knew an open structure was found, we eliminated the second row of the table (closed structure) and used only the cells in the top row to calculate the revised probability.

If you had assumed that the total number of wells was 100 instead of 200, *all* the numbers in the contingency table would be halved. Thus, the conditional probability would still be the same. When you are given the probability data as you were in this example, choose a sample size that keeps the math simple (such as 100).

The scanner problem. A computer chip company uses a scanning device which places a mark on each defective chip it spots in the production line. Ninety percent of all chips are good based upon quality control reports. The remaining 10 percent are defective.

When the chip is okay, the scanner correctly leaves it unmarked 90 percent of the time. When the chip is defective, the scanner correctly marks it as defective 90 percent of the time.

Suppose someone selects one of the chips at random from the production line and tests it on the scanner. The scanner marks the chip as defective. What do you think is the revised probability that the chip is really defective if the scanner marks it as defective? That is, what's the conditional probability that the chip is defective, given that the scanner marked it as defective?

Without doing any math, circle the appropriate percentage.

0% 10% 20% 30% 40% 50% 60% 70% 80% 90% 100%

Now construct a contingency table and compute the probability that a chip is defective given that the scanner marks it as defective, using contingency tables. Then compare your estimate with the correct answer.

The two dimensions of this contingency table are (1) whether the chip is okay or defective and (2) whether the scanner marks it as defective or leaves it unmarked. Let's assume that the probability data are based upon 100 chips. Here's the contingency table data.

100 Chips			
90 Okay Chips		10 Defective Chips	
81 Scanner leaves it unmarked	9 Scanner marks it	1 Scanner leaves it unmarked	9 Scanner marks it

Do you understand how the last row of numbers was determined? Of the good chips, the scanner leaves them unmarked 90 percent of the time. Thus, if there are 90 okay chips, 90 percent of 90 or 81 chips will be left unmarked by the scanner. Given a bad chip, the scanner will mark it as bad 90 percent of the time. If there are ten bad chips, this means that nine chips (90 percent of ten) will be marked as defective. This is the most difficult part of probability revision problems. You may wish to review this and the previous problem before reading on.

Now we are ready to construct the contingency table.

Contingency Table for Scanner Problem

	Scanner Leaves Chip Unmarked	**Scanner Marks Chip As Defective**	**Total**
Chip Is Okay	81	9	90
Chip Is Defective	1	9	10
Total	82	18	100

The probability that a chip is defective given that the scanner marked it as defective is

$$\frac{9}{18} = .50 \text{ or } 50 \text{ percent.}$$

Did you guess about 80 to 90 percent? Most people do. But using a contingency table tells us that you are only justified in revising the probability that you have a bad chip from the original ten percent figure to 50 percent. By now you should be convinced that you need to look at a contingency table to revise probability estimates when you receive additional information.

Let's compute the probability that a chip is defective given that the scanner leaves it unmarked. From the contingency table this is equal to

$$\frac{1}{82} = .012 \text{ or } 1.2 \text{ percent}$$

Thus, if the scanner leaves a chip unmarked it is highly unlikely to be defective. However, if the scanner marks the chip as defective, there is only a 50 percent chance that the chip is really defective.

As we conclude the chapter on basic probability concepts, review the objectives for each unit of the chapter. Have you mastered them? If you haven't, review the material and the exercise sets. Basic probability concepts are at the very core of probability distributions. That's the next chapter.

Exercise Set for 2:6

1. The passing rate in a math class is 70 percent. Of those that passed, 90 percent scored higher than 550 on their math SAT. Of those that failed, 60 percent scored lower than 550 on their math SAT. What is the probability of a student passing given he or she scored higher than 550 on the math SAT? Make an educated guess and then compute the answer. Compare your guess to the answer. (Remember, start with the contingency table).
2. A recent poll revealed that 60 percent of all Republicans are in favor of the school prayer amendment. Only 30 percent of all other voters are in favor of the amendment. Assume that 30 percent of registered voters are Republican. If you select a registered voter at random and the voter is in favor of the amendment, what are the chances that the voter is a Republican? Make an educated guess, then compute the actual answer and compare your guess to the answer.
3. Researchers estimate that one in ten people are creative. They are evaluating a new test to identify who is creative. Of the creative people who have taken the test, 60 percent have scored higher than 70. Of the noncreative people who have taken the test only 5

percent have scored higher than 70. How good is the test? Without the test we know that the chances of selecting a creative person at random is ten percent. What are the chances of selecting a creative person given that he or she scored higher than 70 on our test?

4. A medical researcher wishes to study a new test for detecting hepatitis. He randomly selects hepatitis patients from a hospital. He then selects a group of healthy people who are similar in age and other factors to the hepatitis patients. He gives both groups the detection test. Of those patients that have hepatitis, the test correctly indicates that they have hepatitis 85 percent of the time. Of those who do not have hepatitis, the test indicates that they do not have hepatitis 90 percent of the time. How good is the test? The U.S. Public Health Service reports that only one percent of the population will contract hepatitis. How much should you revise your probability that a person has hepatitis based upon the detection test results? More specifically, what are the chances of having hepatitis given the test says you do? What is the probability of not having hepatitis given the test says you do not? Start by constructing a contingency table. The two dimensions are: (1) whether or not you have hepatitis, and (2) whether the test says you have hepatitis or not.

3. Probability Distributions

3:1 INTRODUCTION

This is a chapter about chances and how to compute them. As you already know, the basic probability expression from Chapter 2 is too restrictive. It assumes that on a single trial, all the outcomes are equally likely. I'll relax that assumption in this chapter and talk about some other ways to calculate probabilities.

Here are some situations for which you might want to compute probabilities. If a bakery has five ovens and needs at least four of them working to meet daily customer demand, what are the chances of meeting customer demand on a given day? What are the chances that all three children in one family will be boys? How effective is a new vaccine and what are the chances of getting the disease if you have been vaccinated?

What is common about these problems? The outcomes on a single trial are *not* equally likely. We will not be able to use the basic probability expression to compute probabilities. Let's examine each situation.

A bakery has five ovens. On any day an oven may be working or not working. Think of whether each oven works or not on a given day as a trial. What are the chances of four ovens working out of five trials? In order to use the basic probability expression to compute the probability, the chances that each oven works or not would have to be equal (50 percent). This is not likely to be the case.

According to the U.S. Public Health Service data, the chances of giving birth to a boy or a girl are not the same. If you wanted to determine the chances of having all boys in a family with three children, the expression from Chapter 2 would not work.

A vaccine may work or not work on a patient (think of a patient as a trial). You wouldn't expect that an effective vaccine would only have a 50 percent chance of working or not working. That would be an unreliable vaccine and the Food and Drug Administration would surely not approve it. Yet unless the chances of working and failing were the same, we have no way yet to compute probabilities. We will shortly—that is the purpose of Chapter 3.

3:2 RANDOM VARIABLES AND PROBABILITY MODELS

By the end of this unit you should be able to:

1. Explain in your own words what a random variable is.
2. Differentiate between discrete and continuous random variables.
3. Distinguish between a probability model, a probability distribution, and a probability histogram.

3:2:1 Random Variables

Before discussing how to compute probabilities where the outcomes on a single trial are not equally likely, we need to distinguish two types of probabilities. Here are some probabilities to use as examples:

1. The probability of having two boys out of three children.
2. The probability of at least three ovens not working on Tuesday.
3. The probability of a computer chip lasting at least 1,000 hours.
4. The probability that a package weighs at least two pounds.

In the first example you are interested in the number of boys out of three children in one family. What's the least number of boys you could have in three children? What's the largest number of boys? What are all the possible values for the event? Please write your answers below.

In the second example you are interested in the number of ovens not working out of five ovens. Please answer the same questions in the space below.

In the third example the event you are interested in is the operational life of a computer chip. Suppose the manufacturer says that chips last between 800 and 1,300 hours. These are the smallest and largest values for the event. What are all the possible operational lifetimes? Are there only a few? How many possible values are there for the lifetime of the chip? Insert your answer below.

In the last example you are interested in the weight of a package. Suppose that several people pick it up and agree that it must weigh more than one pound but cannot weigh more than five pounds. How many different weights could the package be? Is it a small number as in the first two examples? Or is it a very large number of values as in the third example?

Look back over your responses. Here is what you should have.

Event	Number of Possible Values
1. number of boys in three children	four values—0, 1, 2, 3
2. number of ovens not working out of five in the bakery	six values—0, 1, 2, 3, 4, 5
3. lifetime of chip	an *infinite* number of lifetimes between 800 and 1,300 hours. Even rounded to the nearest hour, there would still be a *very large* number of possible values.
4. weight of package	an *infinite* number of weights between one and five pounds. Even rounded to the nearest ounce, there would still be a *very large* number of possible values.

Now you are ready for a formal definition of a random variable.

A **random variable** is an event for which you would like to calculate a probability. The possible number of values for the event must be greater than one. Events that can only be a single value are called constants.

Only four and six values were possible for the first two events. These events are called **discrete random variables**. In the last two situations, an infinite number (or in practice, a very large number) of values were possible. These events are called **continuous random variables**.

3:2:2 Probability Models, Distributions, and Histograms

A probability model helps you calculate probabilities for a random variable. You've already used one probability model—the basic probability expression from Chapter 2. We used it when all outcomes on a single trial were equally likely. If the event didn't meet this *basic assumption* we couldn't compute probabilities this way. There are other models for other situations. Each model has its own assumptions that must be met in order to use the model. In this chapter, two other models are presented: the **binomial probability model** for discrete random variables and the **normal (or bell-shaped) probability model** for continuous random variables.

Let's use the basic probability model from Chapter 2 one more time. We'll compute the probabilities for all values of the *discrete* random variable, the number of boys in a family with three children. To do this the chances of giving birth to a boy or a girl must be equally likely. Here are the four probabilities (if you disagree with the following probabilities you should review 2:2).

The probability of no boys is $^1/_8$.
The probability of one boy is $^3/_8$.
The probability of two boys is $^3/_8$.
The probability of three boys is $^1/_8$.

The four probabilities make up the *probability distribution* for the random variable, the number of boys in a family with three children (when a boy or a girl is equally likely). As you remember from Chapter 2, the sum of the four probabilities must equal one. Here's a formal definition of a discrete probability distribution.

A **discrete probability distribution** consists of all the possible values of a random variable and their associated probabilities.

The same idea applies to continuous probability distributions except that we can't list all possible values of a continuous random variable. I'll have more to say about continuous probability distributions later.

You can describe a discrete probability distribution by listing all the possible values and their probabilities, or by drawing a picture. The picture is called a **probability histogram**. Histograms, as described in Chapter 1, are used to organize raw data. They transform a mass of data into meaningful information. Below is a probability histogram for the probability distribution above.

Figure 3-1 Probability Histogram for the Number of Boys in a Family with Three Children

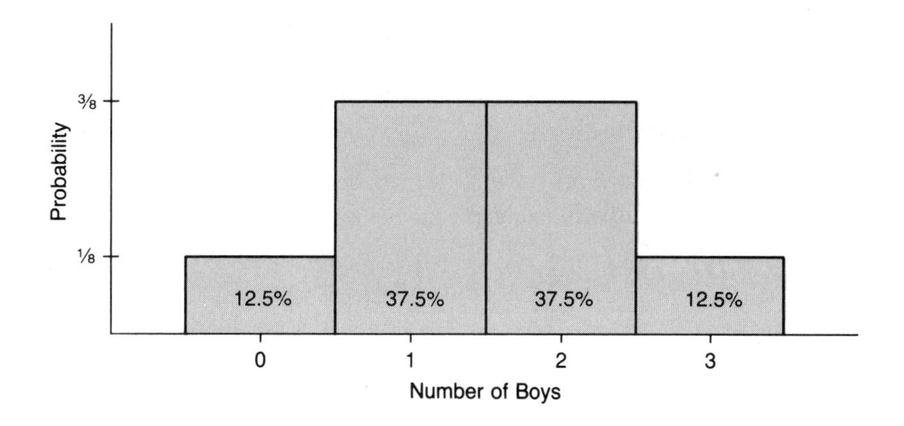

A probability histogram is a new kind of histogram; it is a picture of a probability distribution. Like other histograms, it is made up of bars or rectangles. The base of each rectangle is centered on one of the values of the random variable (for example, zero, one, two, and three boys). The base of the rectangle is one unit wide; its height is equal to the probability of getting that value. The total area under the probability histogram is 100 percent.

In summary, you can compute probabilities for either discrete or continuous random variables using probability models. Each model requires the user to make certain assumptions. The set of all possible values of the random variable and their probabilities is called the probability distribution. A probability histogram is a picture of the probability distribution.

This chapter will present two commonly used probability models. Neither model assumes that on a single trial all the possible outcomes are equally likely. That would be too restrictive for real-world problems, but each model does require the user to make certain assumptions. It was Milton Freidman who said, "there ain't no such thing as a free lunch," so when you use a probability model, you must be sure you are not violating its assumptions. If you do, the probability estimate you compute won't be meaningful or valid.

3:3 A BERNOULLI PROCESS AND STATISTICAL INDEPENDENCE

By the end of this unit you should be able to:

1. Say, in your own words, what a Bernoulli process is.
2. Compute joint probabilities when events are statistically independent.
3. Compute joint probabilities when events are statistically dependent.

3:3:1 The Characteristics of a Bernoulli Process

A Bernoulli process (named after Jacques Bernoulli) has two characteristics.

1. You must be able to classify the outcomes of any trial into two mutually exclusive and exhaustive categories.

a working or nonworking oven
heads or tails for a coin
a boy or girl child
a sick or healthy patient
a defective or acceptable computer chip

Sometimes there are many possible outcomes that can occur on a single trial. If you roll a die there are six possible outcomes that can occur:

1, 2, 3, 4, 5, or 6

If you select voters from the population, three outcomes can occur:

Democrat, Republican, or Independent

In both cases you could still classify the outcomes into two mutually exclusive and exhaustive categories. For example:

a three or a nonthree
a Republican or a non-Republican

2. The probability of an outcome (oven working, boy child, defective chip, etc.) must remain constant from trial to trial.

The second condition is the most important characteristic. It is also more difficult to understand. What the second condition says is that whatever the probability of the outcome is, it *will not be affected* by what happened on preceding trials. Let's explore this idea.

Suppose the probability of getting heads for an unfair or weighted coin is .6. Thus the chance of getting heads on the first flip is .6. What is the chance of getting heads on the second coin flip given you got heads on the first coin flip? Why?

What are the chances of getting heads on the third flip given you had gotten two heads on the first two flips?

What you probably concluded is that the probability of getting heads doesn't change over the three trials no matter what happens. It will always remain .6 because what happens on one trial does not affect what happens on subsequent trials. You have just concluded that you can model the flipping of an unfair coin as a Bernoulli process.

Imagine an urn with five poker chips. Three chips are red and two chips are blue. What is the probability of selecting a red chip from the urn? If you replace it, so the original five chips are back in the urn, what is the probability of selecting a red chip on the second draw given you got a red chip on the first draw? I hope you concluded that the probability was $^3/_5$ in both cases. Since you replaced the chip, the probability of selecting a red chip does not change from trial to trial. Again, this is a Bernoulli process.

Now let's relate the Bernoulli process idea to Chapter 2. For the coin flip experiment with the weighted coin, is the probability of getting heads on the first trial a simple or conditional probability? Why?

It is a simple probability and is equal to .6.

Is the probability of getting heads on the second flip given you got heads on the first flip a simple or conditional probability? Why?

It is a conditional probability because you are looking for the probability of an event *given* that another event preceded it. Earlier you probably said that this probability was also .6. Would you have changed your probability of getting heads on the second flip if the first flip was tails? Are the outcomes on one trial affected by the outcomes on previous trials? No. These are statistically independent outcomes. The conditional probabilities equal each other and also equal the simple probability. No matter what happened on the first trial, it will not change the probability of getting heads on the second trial. Relating this to the previous chapter, you should now be able to conclude:

To have a Bernoulli process the outcomes of successive trials must be *statistically independent*.

Is selecting chips from an urn *with replacement* a statistically independent (or Bernoulli) process? Let's see if the conditional probabilities change from trial to trial.

1. The probability of selecting a red chip on the first draw is $^3/_5$.
2. The probability of selecting a red chip on the second draw given that a red chip was picked on the first draw is still $^3/_5$.
3. The probability of picking a red chip on the second draw after picking a blue chip on the first is still $^3/_5$.

Since the conditional probabilities do not change when you sample *with replacement*, we have statistically independent events and a Bernoulli process.

3:3:2 Computing Joint Probabilities for a Bernoulli Process

A joint probability is the probability of two or more events occurring. Some examples from the last chapter included:

Probability of selecting an employee who is male and promoted
Probability of selecting a voter who is in favor of increased military spending and a Republican
Probability of selecting an employee who is female and not promoted

In these examples, each individual had an equal chance of being selected. The probabilities were all determined using the basic probability model. You can't always use the basic model. In the unfair coin example the two outcomes (heads or tails) were not equally likely. How can you compute joint probabilities when the outcomes on a single trial are not equally likely and you have a Bernoulli process?

You should use the **special rule of multiplication**. This rule says that

the joint probability of two or more independent events is equal to the *product* of their simple probabilities.

For example, the probability of Events A and B, which are statistically independent, is equal to the probability of Event A times the probability of Event B. A shorter way of writing this is:

$$P(A \text{ and } B) = P(A) \times P(B)$$

This rule also applies to more than two events.

$$P(\text{A and B and C and D}) = P(A) \times P(B) \times P(C) \times P(D)$$

Here are some simple examples.

The probability of heads on flip one and tails on flip two (with the weighted coin)

$$P(\text{heads}_1 \text{ and tails}_2) = P(\text{heads}) \times P(\text{tails}) = .6 \times .4 = .24$$

The probability of heads, heads, and tails, in that order

$$\begin{aligned} P(\text{heads}_1 \text{ and heads}_2 \text{ and tails}_3) &= P(\text{heads}) \times P(\text{heads}) \times P(\text{tails}) \\ &= (.6) \times (.6) \times (.4) \\ &= .144 \end{aligned}$$

The special rule of multiplication can be used to compute joint probabilities for statistically *independent* events. What can you do if the events are statistically dependent? There's a more general rule that applies when the events are statistically related. In fact, you'll find that the general rule applies to *all* events, even the special case of independent events. We'll consider statistically related events next.

3:3:3 Characteristics of a Non-Bernoulli Process

You can have a non-Bernoulli process for two reasons. More than two outcomes are possible on a single trial, or the probability of an outcome does not remain constant from trial to trial. While the first violation is trivial, the second problem is fatal. What happens when the probability of an outcome changes from trial to trial? Are these statistically independent events or not? How do you compute joint probabilities?

Let's return to the urn with five poker chips. This time you will select the chips *without* replacement. Is the probability of getting a red chip now affected by previous draws? Suppose you draw three red chips without replacement.

What is the probability of a red chip on draw one?

What is the probability of a red chip on draw two, given that you got a red chip on draw one?

What is the probability of a red chip on draw three after getting red chips on the first two draws?

Please determine the three probabilities and insert your answers in the following space.

Hopefully, you have the following answers:

The probability of a red chip on draw one is $^3/_5$.

The probability of a red chip on draw two, given that you got a red chip on draw one, is $^2/_4$.

The probability of a red chip on draw three after getting red chips on the first two draws is $^1/_3$.

Because you are not replacing the chip, the probability changes after each draw or trial. The two conditional probabilities are not the same. Nor are they the same as the simple probability. This is statistical dependence. If the outcomes were independent, the probability of getting a red chip would not be affected by what happened on previous draws. Thus you should now be able to conclude:

When you have statistically dependent events you don't have a Bernoulli process.

3:3:4 Computing Joint Probabilities for a Non-Bernoulli Process

For statistically dependent events you should use the **general rule of multiplication** for computing joint probabilities. This rule says:

The joint probability of two or more dependent events is equal to the product of their simple and conditional probabilities.

$$P(\text{A and B}) = P(\text{A}) \times P(\text{B } \textit{given} \text{ A})$$

$$P(\text{A and B and C}) = P(\text{A}) \times P(\text{B } \textit{given} \text{ A}) \times P(\text{C } \textit{given} \text{ A and B})$$

Here are some simple examples. Drawing from an urn with three red and two blue chips, without replacement, what are the probabilities of getting

A red chip on the first draw and a blue chip on the second
A blue chip on the first draw and a blue chip on the second
A red chip and two blue chips in that order

$$P(\text{red chip}_1 \text{ and blue chip}_2) = {}^3/_5 \times {}^2/_4 = {}^6/_{20}$$

$$P(\text{blue}_1 \text{ and blue}_2) = {}^2/_5 \times {}^1/_4 = {}^2/_{20}$$

$$P(\text{red}_1 \text{ and blue}_2 \text{ and blue}_3) = {}^3/_5 \times {}^2/_4 \times {}^1/_3 = {}^6/_{60}$$

Let's look at a tree diagram to see how this works when we draw two chips. There is a fork in the tree for each draw, and at each fork there are two branches—a red chip or a blue chip.

Figure 3-2 Tree Diagram: Drawing Two Chips from an Urn Without Replacement

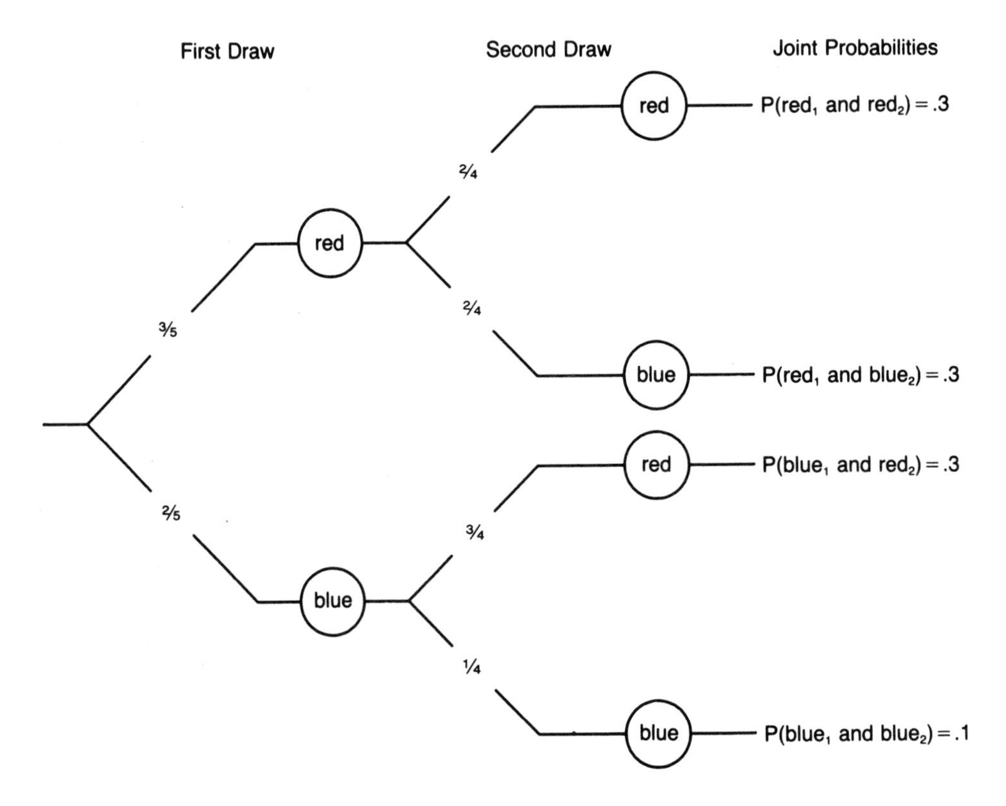

For any given outcome (sequence of two draws) there is only one path. Look at the sequence (red_1, $blue_2$). The probability of being on the red branch after the first draw is $^3/_5$. The conditional probability of getting a blue chip on the second draw given that you got a red chip on the first draw is $^2/_4$. In other words, once you are on the red branch, the probability of moving on to the blue branch is 50 percent. But the *joint* probability of getting on the red branch and then moving on to the blue branch is $^3/_5 \times {}^2/_4 = {}^6/_{20}$ or 30 percent.

Notice that at each fork the probabilities add up to one. At the first fork are the simple probabilities of all possible outcomes on the first draw—a red or a blue chip. Each of the other forks have the conditional probabilities of all possible outcomes *given* the outcome of the first draw that put you on that branch. And, of course, the *joint* probabilities of all outcomes of both draws also add up to one.

What happens when events are statistically independent? The only difference is that the probabilities don't change from trial to trial. The conditional probability of an outcome on the second trial is the same as its simple probability on the first trial, regardless of the outcome on the first trial.

Let's look at a tree diagram for the unfair coin experiment.

Figure 3-3 Tree Diagram for Two Flips of an Unfair Coin

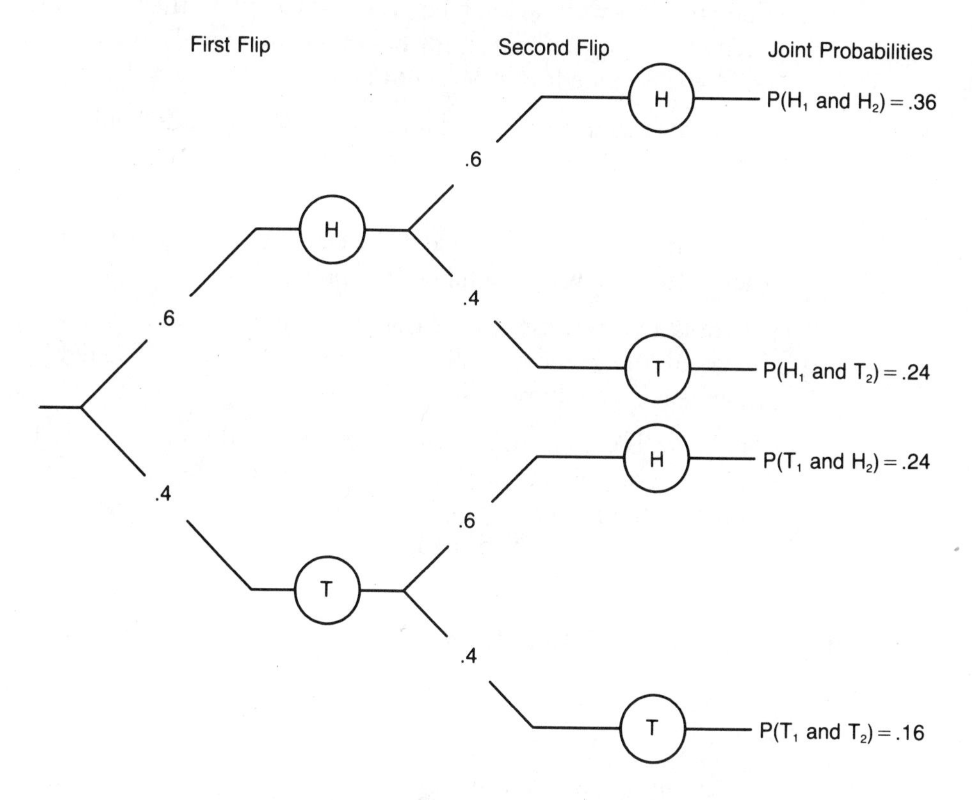

The probability of getting heads is the same at every fork in the tree (.6). So is the probability of getting tails (.4). The conditional probabilities for each outcome are the same as their simple probabilities. So the special rule of multiplication is just a special case of the general rule. When events are independent the conditional probability of an outcome is the same as its simple probability.

If you have a Bernoulli process you should use the special rule of multiplication to calculate probabilities. The binomial model which is presented later in the chapter uses the special rule. If the process is not Bernoulli and the probability of an outcome changes from trial to trial (statistically dependent events), you cannot use the binomial model to calculate meaningful probabilities. There are other models that can be used in cases that don't meet the assumptions of a Bernoulli process. However, these probability models are beyond the scope of this book.

Exercise Set for 3:3

1. The probability of having a boy is .487. Compute the probability of having two boys and a girl, in that order, in a family of three children. Assume it is a Bernoulli process.
2. Draw three cards from a deck of regular playing cards, replacing the card and shuffling between draws. Compute the probability of drawing a heart, a heart, and a nonheart, in that order.
3. Would it make any difference if the card was not replaced after each selection? Why? Compute the probability.
4. There is a 90 percent chance that an oven works on any given day. What are the chances of all five ovens working next Tuesday? Assume a Bernoulli process.
5. There are 100 chips in an urn. Seventy are red (r) and 30 are blue (b). You draw ten chips at random and without replacement. Compute the probability of drawing the following sequence of chips: r, r, r, r, r, r, b, b, b, and b.

3:4 MORE ON A BERNOULLI PROCESS

By the end of this unit you should be able to:

1. Determine either by using logic or analyzing the data whether a problem can be represented as a Bernoulli process or not.
2. Explain to others how you arrived at your conclusion.

It is not always easy to determine if a problem is a Bernoulli process. Few real-world examples are as simple as flipping coins or selecting poker chips out of an urn. So before calculating probabilities, you need to have some additional practice in determining if a problem can be modeled as a Bernoulli process.

Here are two problem situations.

Situation 1. An oil company has options on three tracts of land. One is in Alaska, one is in the North Sea, and one is in Nigeria. The company is interested in calculating the chances of finding oil on all three tracts. The chances of finding oil on any single tract, based on published worldwide data, is .3.

Situation 2. Another oil company owns three tracts of land in a Texas oil field. The tracts are next to one another. The overall chances of finding oil on a single tract is .3. You wish to compute the chances of finding oil on all three of the tracts.

Although not critical, the first condition for a Bernoulli process is satisfied in both situations. On a single trial only two outcomes can occur—oil or no oil.

In one of the scenarios, the second condition of statistically independent events is either exactly or nearly met. In the other, the second condition is violated. Remember, a Bernoulli process and statistical independence mean the probability of an outcome remains constant from one trial to the next. When events are statistically dependent, the probability of an outcome changes from one trial to the next. With these ideas in mind, please determine which scenario can and which cannot be modeled as a Bernoulli process and why. You may use *logic* or *analysis* to make your determination. This is a difficult question so take your time or ask your friends to help.

Here are the correct answers:

1. You can represent Situation 1 as a Bernoulli process. Suppose you strike oil in the Alaskan tract. Should you increase your chances that you will find oil in the North Sea or in Nigeria? Since the oil fields are located throughout the world and have different geologic structures, you should not revise your .3 probability of finding oil at the other sites if you strike oil at one site. Through logic you should conclude that you have statistically independent events and a Bernoulli process.
2. You cannot represent Situation 2 as a Bernoulli process. If you strike oil on one tract of land, you should increase the chances of finding oil on the other two tracts. Remember, the three tracts are near one another and have the same geologic structure. If the chance of finding oil anywhere is .3, but you find oil on Tract 1, the chances of finding oil on Tract 2 would be greater than .3. Again using logic you can conclude that you do not have a Bernoulli process.

Sometimes logic is all you have to go on. But whenever possible, and especially if you have data available, you should avoid using logic alone to determine if you have a Bernoulli process. Analysis usually gives more reliable results. In the following example you can analyze the results of an experiment to see if the problem can be modeled as a Bernoulli process.

The Sex of a Child—A Bernoulli Process? Can the sex of a child be modeled as a Bernoulli process? Does the probability of having your third boy increase if you already have two boys? Does the probability of having your first boy drop if you already have three girls? This hypothetical experiment attempts to answer these questions.

Our subjects were 7,000 pregnant women. They were selected and assigned to seven groups based on the sex of their previous children. Only women who had no previous children or one to three children of the same sex were selected. When all the women gave birth from the latest pregnancy, we recorded the number of boys and girls. Here are the data.

Number and Sex of Previous Children

	None	1 Boy	2 Boys	3 Boys	1 Girl	2 Girls	3 Girls
Latest Child							
Boy	487	488	500	600	480	460	380
Girl	513	512	500	400	520	540	620
Totals	1,000	1,000	1,000	1,000	1,000	1,000	1,000

Do we have a Bernoulli process? Why or why not?

If you still haven't figured it out, these three probabilities should help you.

The probability of having a boy, with no previous children, is

$$\frac{487}{1{,}000} = .487$$

The probability of having a boy, given three previous girls, is

$$\frac{380}{1{,}000} = .38$$

The probability of having a boy given three previous boys is

$$\frac{600}{1{,}000} = .60$$

Do we have a Bernoulli process? *No*, the conditional probabilities change *significantly*. Thus, if this were an actual study, you would not want to use the binomial model to calculate probabilities for the number of boys or girls in a family.

One of the most difficult decisions facing an analyst is deciding whether a problem can be represented as a Bernoulli process. If it can be, you use the binomial model. If it can't, you can't use the binomial model.

Before starting the next unit, take a minute to review the objectives for this unit. If you can't explain to others how you decide whether you can model a problem as a Bernoulli process, please reread the unit.

3:5 THE BINOMIAL MODEL

By the end of this unit you should be able to:

1. Estimate binomial probabilities using simulation.
2. Solve binomial problems without using any formulas—just the idea of statistical independence and the special rule of multiplication.
3. Draw and interpret a probability histogram.
4. Compute and interpret the expected value and standard deviation.
5. Make managerial decisions as a result of the binomial model.

Probability models are used to calculate probabilities for random variables. You use the **binomial model** to calculate probabilities for a discrete random variable—the number of times a given outcome will occur in a sequence of trials—when you have a Bernoulli process. You assume that the probability of the outcome you are interested in remains constant from trial to trial.

There are several ways to find binomial probabilities. You can plug numbers into the binomial formula or look the results up in a table. However, we won't use the formula or tables here because I think they get in the way of understanding. Instead I'll show you how to compute binomial probabilities just by using the special rule of multiplication. Another way is to use simulation. You will only be able to *estimate* the exact probability using simulation. However, it is a simple way, it is not mathematical, and you are unlikely to forget it.

3:5:1 Estimating Binomial Probabilities by Simulation

The word "simulation" means the act of performing small-scale experiments using a table of random digits. Every statistics book has a table of random digits. While the digits are generated using a very sophisticated computer program, here's how you could generate them manually. Imagine a huge barrel with billions of single digit chips numbered from zero to nine. There are equal numbers of each digit in the barrel. The barrel is constantly rotated which mixes the chips up thoroughly. You select one chip at a time from the barrel. If you continued to do this and recorded the digit number of the chip you would have a table of random digits.

A table of random numbers is a list of the ten digits 0, 1, 2, 3, 4, 5, 6, 7, 8, and 9 which has the following properties.

1. The sequence of single digits in the table exhibits no detectable pattern.

2. All the digits have the same chance of being anywhere in the table.

Here are 144 random digits. The digits are clustered in pairs to make them easier to read.

A Very Small Table of Random Digits:

03	97	16	12	55	16	84	63	33	57	18	25	23	52	37	70	56	99
16	31	68	74	27	00	29	16	11	35	38	31	66	14	68	20	64	05
07	68	26	14	47	74	76	56	59	22	42	01	21	60	18	62	42	36
85	22	29	69	49	08	16	34	57	42	39	94	90	27	24	23	96	67

Let's return to the bakery problem. A bakery has five ovens. Four of the ovens must be working in order to meet customer demand on any given day. The owner keeps records for major repairs. A major repair is one that requires a service call and puts an oven out of service for the day. A minor repair, such as a blown fuse or sticky oven door, takes only a few minutes to fix and there are no lost sales.

The owner has kept major repair records over the last ten years. She has determined that the chance of an oven needing a major repair on a given day is .10, and that oven breakdowns are statistically independent. That is, one oven breaking down does not affect the probability that the others will work. The random variable we are interested in is "the number of ovens working on any given day." The probability is .90 (remember the chance that an oven needs a major repair is .10).

We will use simulation to *estimate* the probability of at least four ovens working on a given day. This is a Bernoulli process, so later we'll calculate the *exact* probability using the binomial model. This will give us a chance to compare the simulation technique with the more mathematical approach.

We will use a single digit to simulate, or represent, whether an oven works or not on a given day.

The following digits	represent	with a probability of
0, 1, 2, 3, 4, 5, 6, 7, 8	oven is working	.9
9	oven is not working	.1

The digit "9" represents an oven not working, but any digit could have been used.

Let's simulate 25 days of operations. Five random digits are drawn (because there are five ovens in the bakery) from the table to simulate the operations for each day. If a given digit is between zero and eight, that will represent a working oven. The digit nine will simulate (or stand for) an oven which is not working. Random digits were selected from the beginning of the table. The results of the simulation are shown below.

Day	Random Digits	Interpretation
1	03971	Four ovens working and one not working.
2	61255	All five working.
3	16846	All five working.
4	33357	All five working.
5	18252	All five working.
6	35237	All five working.
7	70569	Four ovens working and one not working.
8	91631	Four ovens working and one not working.
9	68742	All five working.
10	70029	Four ovens working and one not working.
11	16113	All five ovens working.
12	53831	All five ovens working.
13	66146	All five ovens working.
14	82064	All five ovens working.
15	05076	All five ovens working.
16	82614	All five ovens working.
17	47747	All five ovens working.
18	65659	Four ovens working and one not working.
19	22420	Five ovens working.
20	12160	Five ovens working.
21	18624	Five ovens working.
22	23685	Five ovens working.
23	22296	Four ovens working and one not working.
24	94908	Three ovens working and two not working.
25	16345	All five ovens working.

In 24 of the 25 days at least four ovens were working (on day 24 less than four ovens worked). By simulation:

$$P(\text{at least four ovens working}) = 24/25 = .96$$

If the probability of an oven working was .84, how could we simulate that event? This time we must use pairs of digits.

The following digits	represent	with a probability of
00, 01, 02, 03, . . . 83	oven working	.84
84, 85, 86, 87, . . . 99	oven not working	.16

The one weakness of simulation is that it only provides an estimate of the exact probability. The larger the experiment (in this example—adding more days) the better will be the estimate.

Here's another example for simulation. You wish to compute probabilities for the following random variable—the number of girls in families with three children. Assume that the chance of having a girl is .6 and is a Bernoulli process. Estimate the probability of having no boys in a family of three children. Without looking ahead, how would you assign the random digits?

The following digits	**represent**	**with a probability of**
0, 1, 2, 3, 4, 5	having a girl	.6
6, 7, 8, 9	having a boy	.4

Simulate twenty families with three children. Take three numbers at a time from the table of random digits. If none of the three digits are higher than five, all three children are girls in that simulated family. Do this twenty times starting at the beginning of the table.

Starting at the beginning of the table, four out of the first twenty groups of three digits have no digits greater than five. The estimated probability is $^4/_{20}$ or .2. As you will see in the next section, that is a good estimate. Of course, if we increased the experiment from 20 families to 200 or, better yet, 2,000 families with three children, we would get a more accurate estimate. If you plan to run large-scale simulations, you should have a computer select the random digits for you.

3:5:2 Computing Binomial Probabilities

Let's return to the previous example in which the probability of having a son is .4. The random variable—the number of boys in families with three children—can take on only four values, zero, one, two, or three boys. What are the chances of having two boys and one girl? Here is a very simple four-step procedure you should use for calculating exact probabilities. It is based upon the special rule of multiplication.

1. Arrange the outcomes that you are interested in in any sequence.

We are interested in computing the probability of two boys(B) and one girl(G). One possible sequence is B, B, and G.

2. Compute the probability of obtaining the sequence using the special rule of multiplication.

Because we have a Bernoulli process, the probability of having a boy and a boy and girl is

$$P(\text{boy and boy and girl}) = P(\text{boy}) \times P(\text{boy}) \times P(\text{girl})$$
$$= .4 \times .4 \times .6 = .096$$

3. Determine the number of different ways you can arrange the outcomes you are interested in—two boys and one girl.

We are interested in determining the chances of having two boys and a girl—not in any special sequence. The B, B and G sequence is only one of several possible sequences. How many sequences are there? You should be able to list the three different sequences that include two boys and one girl.

BBG, BGB, and GBB

Each sequence has the same probability. Check it and see.

$$P(\text{B and B and G}) = .4 \times .4 \times .6 = .096$$
$$P(\text{B and G and B}) = .4 \times .6 \times .4 = .096$$
$$P(\text{G and B and B}) = .6 \times .4 \times .4 = .096$$

4. Multiply the probability of one sequence by the number of different sequences. The probability of having two boys and one girl is:

$$3 \times .096 = .288 \text{ or } 28.8 \text{ percent of all families with three children.}$$

In summary, this is the procedure:

1. Arrange the outcomes that you are interested in in any sequence.
2. Compute the probability of obtaining the sequence using the special rule of multiplication.
3. Determine the number of different ways you can arrange the outcomes you are interested in.
4. Multiply the probability of one sequence by the number of different sequences.

When the problem is small it is easy to determine the number of different sequences. All you do is list them and count. But when you are dealing with many possible events, it's very time consuming. There is a much faster way which is called **combinations**. The number of different sequences can be determined by the following expression.

$$\frac{(\text{total number of outcomes})!}{(\text{number of boys})! \times (\text{number of girls})!}$$

The symbol "!" stands for the factorial. To compute the factorial, you start with the number before the factorial symbol and multiply it by the next lower integer. Multiply this product by the next lower integer, and so on. Continue to do this until you come to one. For example, 3! equals 3 × 2 × 1 or 6. Zero factorial is equal to one. (Don't ask why unless you want a very mathematical answer.)

In our example the total number of outcomes is three. Two outcomes are boys (2!) and the other one must be a girl (1!). Thus we have:

$$\frac{3!}{2! \times 1!} = 3$$

Try the four-step procedure and compute the following three probabilities: P(no boys), P(one boy), P(three boys). Please don't look ahead.

Here's how to compute the probability of having one boy.

Step 1: One sequence is b, g, and g

Step 2: By the special rule of multiplication

$$P(\text{b and g and g}) = .4 \times .6 \times .6 = .144$$

Step 3: How many different sequences are there?

$$\frac{3!}{(1\ \text{boy})! \times (2\ \text{girls})!} = 3$$

Step 4: The probability of having one boy and two girls (in any order) is

$$3 \times .144 = .432$$

or 43.2 percent of all families with three children

If you made no math errors, here is what you should have:

$$P(\text{no boys}) = .216$$

$$P(1\text{ boy}) = .432$$

$$P(2\text{ boys}) = .288$$

$$P(3\text{ boys}) = .064$$

There are two important observations to make. First, the estimated probability of having no boys (or all girls) from the simulation was .20. The exact probability is .216.

Second, since all the possible numbers of boys are accounted for, the sum of the four probabilities *must* equal one. They do. In fact, we have just computed a probability distribution for the random variable—the number of boys in families with three children (when the probability of having a boy is .4). Now we can draw a probability histogram for the random variable.

Figure 3-4 Probability Histogram: Number of Boys in a Family with Three Children When the Probability of a Boy is .4

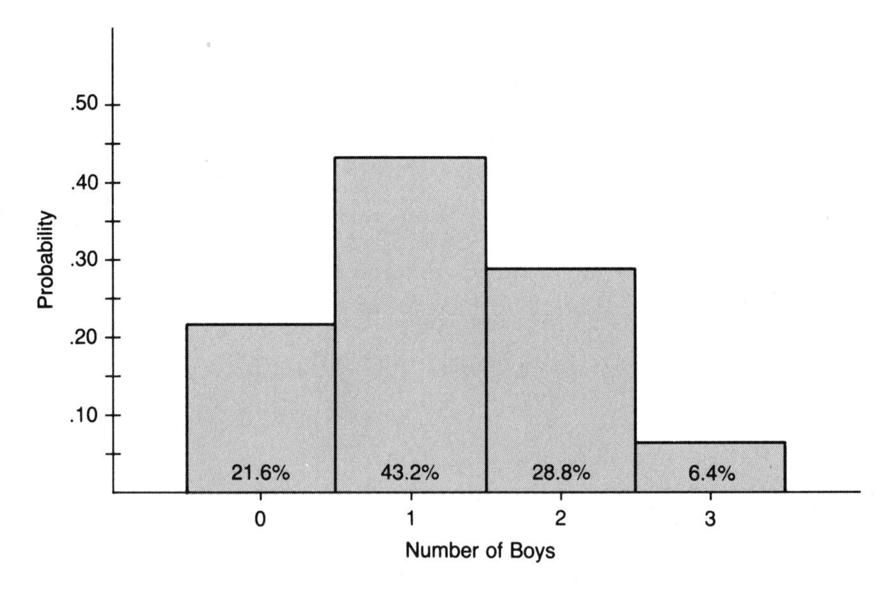

Let's do another problem. Return to the urn problem from Chapter 2. We had ten urns each with 100 poker chips. Six of the urns contained 70 blue and 30 red chips. Four contained 70 red and 30 blue chips. I selected one urn at random and the chances that it contained 70 red and 30 blue chips was .4. Then I selected ten chips *with replacement* from the urn and got six red and four blue chips. Now what are the chances that the urn I selected contained 70 red and 30 blue chips?

In Chapter 2, I argued that six red and four blue chips were much more likely to come from a 70 red, 30 blue urn than from a 70 blue, 30 red urn. Chip selection *with replacement* is a Bernoulli process. The probability of getting a red chip does not change as you select ten chips from the urn. Now we need to compute the following two probabilities.

1. What are the chances that we would have gotten six red and four blue chips from an urn that contained 70 red and 30 blue chips?
2. What are the chances that we would have gotten six red and four blue chips from an urn that contained 70 blue and 30 red chips?

Take a few minutes and use the four-step procedure to solve these two problems.

1. From an urn that contained 70 red and 30 blue chips:

Step 1: One sequence is r, r, r, r, r, r, b, b, b, and b

$$P(\text{red}) = .7 \text{ and } P(\text{blue}) = .3$$

Step 2: The joint probability of this sequence is

$$.7 \times .7 \times .7 \times .7 \times .7 \times .7 \times .3 \times .3 \times .3 \times .3 = .000953$$

Step 3: The number of possible different sequences is

$$\frac{10!}{(6 \text{ red})! \times (4 \text{ blue})!} = 210$$

Step 4: The probability is 210 × .000953 or .2001

2. From an urn that contained 70 blue and 30 red chips:

Step 1: One sequence is r, r, r, r, r, r, b, b, b, and b

$$P(\text{red}) = .3 \text{ and } P(\text{blue}) = .7$$

Step 2: The joint probability of this sequence is

$$.3 \times .3 \times .3 \times .3 \times .3 \times .3 \times .7 \times .7 \times .7 \times .7 = .000175$$

Step 3: The number of possible different sequences is the same as above, 210

Step 4: The probability is 210 × .000175 or .0368

You are more likely to get six red and four blue chips from an urn with 70 red and 30 blue than from an urn with 70 blue and 30 red. That is why in Chapter 2 the probability was increased from .40 to about .80.

3:5:3 Expected Value and Standard Deviation of the Binomial Probability Distribution

In Chapter 2 we discussed the idea of the long run average, or expected value. We can get a quick estimate of expected value by "eyeballing" the balance point of the probability histogram. Here's the probability histogram for the random variable, the number of boys in families with three children, when the probability of a boy is .4.

Figure 3-5 Number of Boys in Families with Three Children

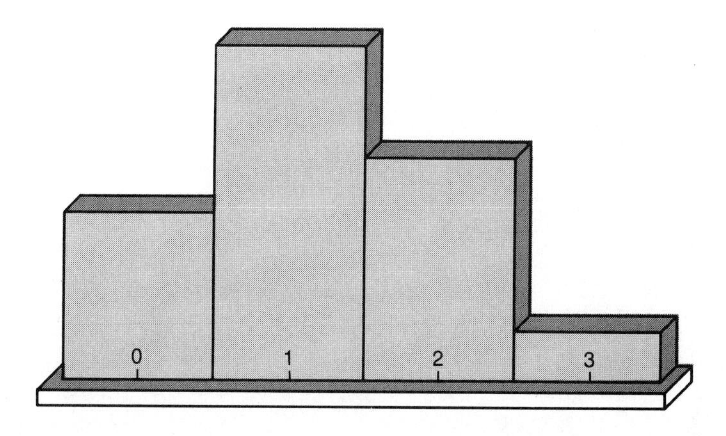

Imagine moving the histogram back and forth on a steel rod until you get the histogram to balance on the rod. At what point would it balance?

Hopefully, you concluded that the histogram would balance somewhere between a value of one and two, probably closer to the value of one. This means that families with three children, on the average, have somewhat more than one boy.

Let's compute the expected value and see how close your estimate is. Expected value is a weighted average. It is the sum of all the possible outcomes weighted by their likelihoods of occurring.

$$\begin{aligned} \text{expected value} &= (0 \text{ boys} \times .216) + (1 \text{ boy} \times .432) \\ &\quad + (2 \text{ boys} \times .288) + (3 \text{ boys} \times .064) \\ &= 1.2 \text{ boys} \end{aligned}$$

This agrees with the quick estimate. Does that mean that each family with three children will have exactly 1.2 boys? Please give two arguments as to why this can't be true.

First, it's impossible to have exactly 1.2 boys. The second reason is much more important. The expected value is a long run idea. If we kept track of many families with three children, we would find that in the long run, there were 1.2 boys per three children. Expected value is always a meaningful figure for the long run. Under what conditions is it a meaningful number in the short run?

Expected value is also meaningful in the short run when the standard deviation is small. The standard deviation measures the spread of the probability data around its mean or expected value. If the standard deviation was very small, then most families with three children would have very close to 1.2 boys (that is, one or two) and very few would have either no boys or three boys. If the standard deviation was large, then families with three children would tend to have either no boys or three boys.

We cannot postpone computing the standard deviation any longer. Drawing on Chapter 2, the variance for probability data is the squared differences of all the values of the random variable around the expected value weighted by their likelihoods of occurring.

$$\text{variance} = (0 - 1.2)^2(.216) + (1 - 1.2)^2(.432) + (2 - 1.2)^2(.288) + (3 - 1.2)^2(.064)$$

$$= .72$$

standard deviation = .85

In this example the expected number of boys in families with three children is 1.2 and the standard deviation is .85 boys. Since the standard deviation is relatively small, most of the families will have one or two boys.

3:5:4 The Binomial Model and Decision-Making

In this section you will see how the binomial model is used to help make decisions. The key is interpreting the probabilities that you have computed.

To Purchase or Not to Purchase a Sixth Oven? We return to the bakery for the last time. The bakery has five ovens. Based upon the owner's records, the chance of an oven operating on any day is .9. The owner has also observed that as long as four ovens are working, customer demand can be met and no sales are lost (although the bakers' tempers tend to rise along with the dough). But if only three ovens are working, lost sales reduce the bakery's profits by an average of $100 each day. Besides this loss, some disappointed customers may be lost for good.

The bakery can arrange for the purchase of a sixth oven at a cost of $175 per month. The owner has decided that if the new oven can just pay for itself by avoiding lost sales, it will be worth it, since customers and employees will both be happier. But how often can we expect a sixth oven to prevent lost sales? Is this a wise purchase?

In order to determine how much money will be saved we have to know how often sales are lost now with five ovens and how often they would be lost with six ovens. The owner has observed that oven breakdowns are statistically independent so we have a Bernoulli process.

With five ovens, the probability that *at least* four ovens are working is the sum of the probability that four ovens are working plus the probability that five ovens are working. Remember to use the four-step procedure and the special rule of multiplication.

$$\begin{aligned}
\text{P(at least four ovens working)} &= \text{P(four ovens working)} \\
&\quad + \text{P(five ovens working)} \\
&= \frac{5!}{4! \times 1!} \times .9 \times .9 \times .9 \times .9 \times .1 \\
&\quad + \frac{5!}{5! \times 0!} \times .9 \times .9 \times .9 \times .9 \times .9 \\
&= .919 \text{ or } 91.9 \text{ percent of the time}
\end{aligned}$$

This means that in a 30 day month at least four ovens will be working an average of 27.57 days (.919 × 30 days). We will not meet customer demand on an average of 2.43 days. This means that oven breakdowns cause an average of $243 in lost profits each month with five ovens.

If the bakery buys the sixth oven, the probability that *at least* four ovens will be working is the sum of the probabilities that four, five or six ovens will be working.

$$\begin{aligned}
\text{P(at least four ovens working)} &= \text{P(four)} + \text{P(five)} + \text{P(six)} \\
&= \frac{6!}{4! \times 2!} \times .9 \times .9 \times .9 \times .9 \times .1 \times .1 \\
&\quad + \frac{6!}{5! \times 1!} \times .9 \times .9 \times .9 \times .9 \times .9 \times .1 \\
&\quad + \frac{6!}{6! \times 0!} \times .9 \times .9 \times .9 \times .9 \times .9 \times .9 \\
&= .984 \text{ or } 98.4 \text{ percent of the time}
\end{aligned}$$

With a sixth oven we would expect at least four ovens to be working an average of 29.52 days a month (.984 × 30 days). We will not meet customer demand an average of .48 days a month, costing an average of $48 a month. The sixth oven will save an average of $243 − $48 = $195 in lost profits due to lost sales each month. At a cost of $175 per month, the sixth oven looks like a good buy.

Earlier we simulated the chances of having at least four ovens working in a bakery with five ovens. We simulated 25 days of bakery operations. The estimated probability was .96. The exact probability was .919. Not a bad estimate but it may not have been precise enough in a real-world problem. It's best to calculate probabilities rather than estimate them.

Problem Detection and the Binomial. Before you can solve a problem, you have to be aware of it. Sometimes you can use the binomial model to detect a problem before others who have had no training in probability.

A stationary wholesaler knows from past data that 40 percent of her orders will be for more than 100 boxes of paper (big order). Of the last six orders to arrive, none have been for more than 100 boxes. Could the wholesaler be facing a problem with declining sales?

First assume that sales are not declining (the probability is still 40 percent). Shortly we'll compute the chances of having no big orders out of the last six. Suppose the probability turns out to be very small. This means that it would be "a cold day in July" before she would expect to receive no big orders. But in fact she has received no big orders out of the last six. She may be facing a serious decline in her sales.

Let's assume that orders arriving at the wholesaler can be represented by a Bernoulli process. This means that

1. The probability of a big order is .4
2. The probability that the second order is a big order given that the first order was a big order is still .4; and
3. The probability of a third big order given two big orders is still .4.

Using the four-step procedure, the probability of no big orders in the last six orders is

$$\frac{6!}{6! \times 0!} \times .6 \times .6 \times .6 \times .6 \times .6 \times .6 = .047$$

or 4.7 percent of the time.

The chances of getting no big orders out of six is only 4.7 percent. Yet that's what actually happened. What could account for it? There are three possible explanations.

1. The probability of getting a big order has not changed and there is no problem with declining sales. We know that getting no big orders out of six will happen occasionally—4.7 out of 100 times. This is one of those rare times.
2. The probability of getting a big order has changed and we have a problem of declining sales. How do we know we have a problem?

We are more likely to get no big orders out of six, if the chances of getting a single big order dropped from the historic 40 percent figure. For example, if the chance of getting a big order has dropped from 40 percent to 10 percent, the chance of getting no big orders in six would have been 53.1 percent instead of 4.7 percent (check it out).

3. The arrival of orders can no longer be modeled by a Bernoulli process. If this is true, then the 4.7 percent figure is no longer meaningful and cannot be used to detect the presence of a sales problem. We can't tell if we have a problem with declining sales.

If you believed that order arrivals could still be modeled by a Bernoulli process, then you have to decide whether there is a problem with declining sales. We know that only 4.7 percent of the time we would have gotten no big orders if the probability was still 40 percent. Is it one of those rare times? There is no correct answer; you must decide. Many people would feel that this was not one of the 4.7 times in a hundred and would look for causes in the declining sales. Of course, they might turn out to be wrong. What would you do?

It's time to summarize. You should now be able to determine if a problem can be modeled by a Bernoulli process. If it can, you should be able to determine the binomial probabilities either by simulation or the special rule of multiplication and combinations (the binomial model). If the problems cannot be represented by a Bernoulli process, you will have to use probability models which are beyond the scope of this book. You should also be able to draw a probability histogram for a probability distribution, estimate its expected value (the balance point idea), and compute its expected value and standard deviation. Finally you should be able to interpret probabilities so as to solve problems and make correct decisions.

Exercise Set for 3:5

1. A department store has classified its charge customers as either high or low volume purchasers. Twenty percent are high volume purchasers. If a sample of four people are randomly selected from the list of charge customers, what is the probability that none of them are high volume purchasers?
2. Did you need to assume that the chances of selecting a high or low volume purchaser were the same? And if not, why not?

3. Compute the probabilities that there will be one high volume purchaser, two high volume purchasers, three high volume purchasers, and four high volume purchasers, out of a sample of four.
4. Draw a probability histogram. Use the balance point idea to estimate the expected number of high volume purchasers in a sample of four charge customers.
5. Compute the expected number and standard deviation in the number of high volume purchasers out of four charge customers.
6. You plan to drill for oil on three tracts of land. One is in Texas, the second is in Mexico, and the third is in Alaska. The chance of striking oil, based on worldwide statistics is .2. What are the chances of no oil strikes, one oil strike, two oil strikes, and three oil strikes?
7. Draw a probability histogram for Exercise 6.
8. Compute the expected number of oil strikes and the standard deviation.

3:6 CONTINUOUS RANDOM VARIABLES AND THE NORMAL MODEL

By the end of this unit you should be able to:

1. Differentiate between the binomial and the normal models.
2. Compute normal probabilities.
3. Use normal probabilities in problem solving and decision-making.
4. Explain to others why the mean and standard deviation of the normal curve are not known with certainty except in statistics books.

A probability model helps you calculate probabilites for a random variable. We have used the binomial model to compute probabilities for a discrete random variable, when we could model the problem as a Bernoulli process.

Unlike discrete random variables, continuous random variables can take on an infinite number of values. Practically speaking, we consider a random variable to be continuous when it can take on a very large but finite number of values. The most common probability model for calculating probabilities for a continuous random variable is the **normal model.**

3:6:1 Why Is the Normal Model so Useful?

The normal model is useful for a number of reasons. First, it can serve as an approximation for other probability models that require a lot of computation to obtain the exact probability.

You can use the normal model to approximate the binomial model. This is very useful when the discrete random variable can take on a large number of possible values.

This is illustrated in the following three examples. First let's toss a fair coin ten times. The discrete random variable—the number of heads in ten flips, can take on eleven values (0, 1, 2, 3, . . . 10). All 11 probabilities (the probability distribution) have been computed and drawn in the probability histogram below.

Figure 3-6 Probability Histogram for Number of Heads in Ten Tosses

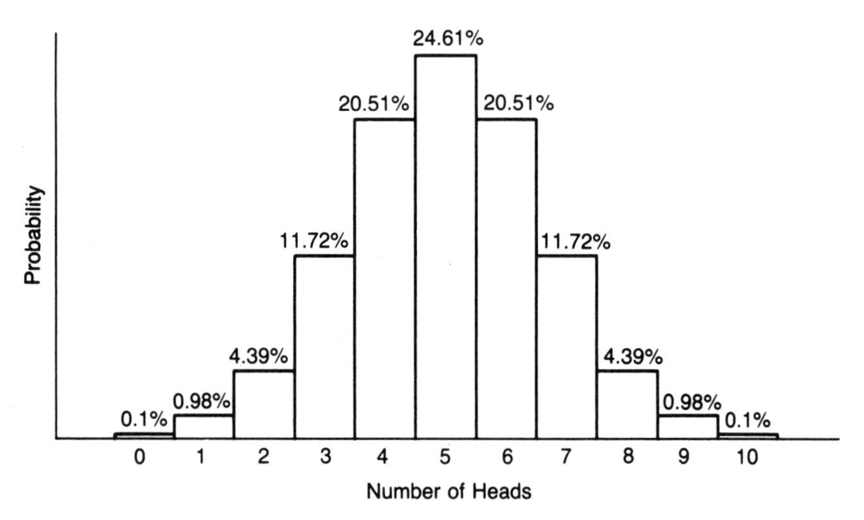

Computing 11 binomial probabilities is not a time-consuming chore. As always, the area underneath the probability histogram is 100 percent.

Now toss the fair coin 100 times. The discrete random variable can now take on 101 possible values (0, 1, 2, 3,. . .100). Computing all 101 probabilities—the probability distribution—would be time-consuming. See the probability histogram below.

Figure 3-7 Probability Histogram for Number of Heads in 100 Tosses

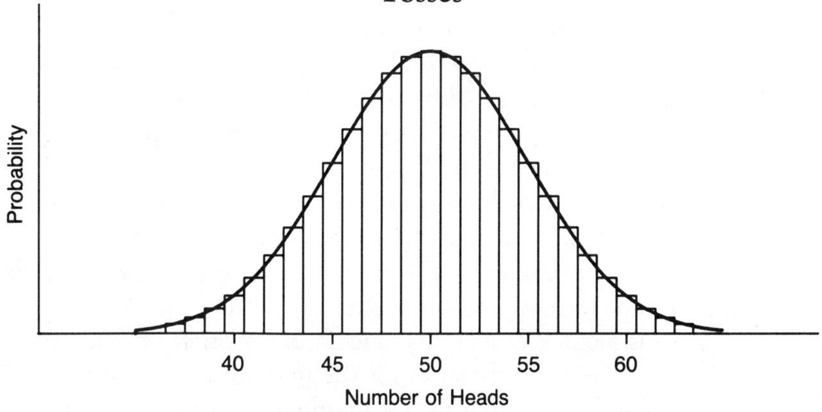

Probability

40 45 50 55 60

Number of Heads

I've drawn a smooth curve through the center of the top of each rectangle. The probability histogram for 100 tosses is already very close to the normal-shaped curve. The area underneath the smooth curve is, of course, 100 percent.

Now we'll toss the coin 999 times. The discrete random variable can now take on 1,000 possible values.

Figure 3-8 Probability Histogram for Number of Heads in 999 Tosses

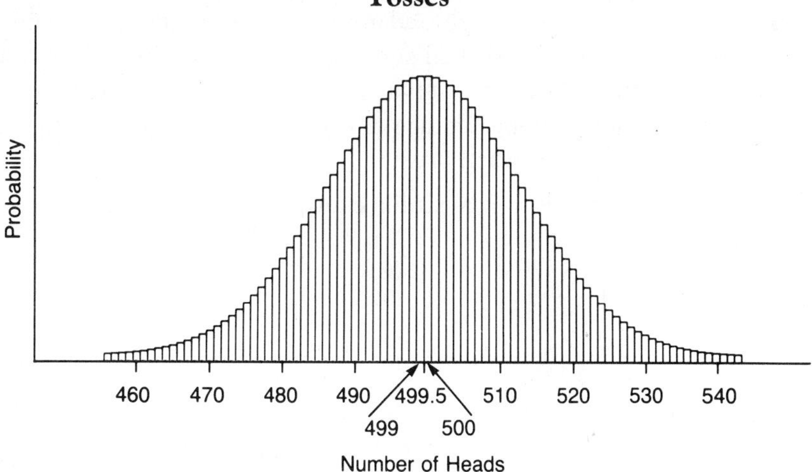

As the number of tosses goes up, the probability histogram follows the normal-shaped curve better and better. For 999 tosses the histogram appears to be a smooth curve. The highest point is exactly in the center, at 499.5 heads. The portion of the histogram to the left of the center is exactly 50 percent of the area, and the probability of getting any value to the left of the center, from 0 to 499, is .50 or 50 percent. Likewise, the probability of getting 500 or more heads (anywhere to the right of center) is 50 percent. And the probability of getting within any interval, say from 490 to 499 heads, is equal to the percentage of the total area that lies between those points. The normal model gives us a way of calculating those areas and probabilities.

If you wanted to compute the probability of getting between 490 and 499 heads in 999 tosses you have two choices: (1) treat the random variable as discrete and use the binomial model or (2) treat the random variable as continuous and use the normal model. If you use the binomial model you will have to compute factorials such as 999!. The latter approach is a lot less work. The normal model can be used as a quick probability calculator for other probability models as well.

A second reason for the importance of the normal model is that it can be used to *approximate* the distributions of many kinds of variables. Variables such as IQ, height of people, grade point average at a university, systolic blood pressure, accidental death rates, and the diameter of machined parts are affected by a large number of factors. When many independent factors affect a variable we often find that the behavior of the continuous random variable can be described by the normal model. For example, systolic blood pressure is influenced by genetic factors, diet, weight, life-style, aerobic conditioning, and other factors. If you sampled 94 people, recorded their systolic blood pressures, and plotted a frequency histogram, it would look *almost* like the normal-shaped curve. (Remember, a frequency histogram shows the *number* of observations in each class, while a probability histogram shows the percentage of time, or the probability, that an observation will fall in each class.)

Figure 3-9 Systolic Blood Pressures of 94 Randomly Selected Males

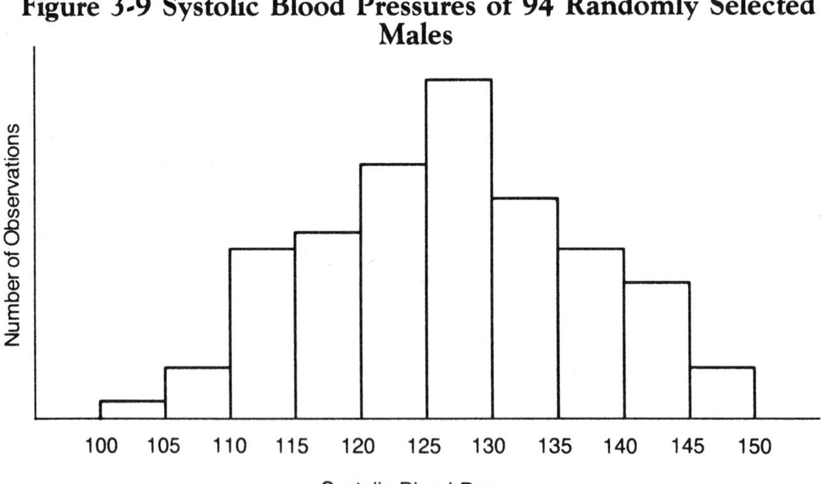

No frequency histogram is exactly normal-shaped, but many are remarkably close. And because they are, we can use the normal model to compute probabilities.

Perhaps the most important reason for the usefulness of the normal model is called the *central limit theorem.* It says that sampling distributions (which we'll discuss in the next chapter) tend to be normal-shaped. What the theorem is and why it is important must wait for the next chapter. But if the central limit theorem didn't exist, statistics would have little practical use. (Even as you read this you are probably thinking this is true.)

3:6:2 Features of the Normal Curve

The probability histograms for the coin toss experiments in Figures 3-6 to 3-8 represent discrete probability distributions. But the smooth curves approximate continuous probability distributions—the probabilities that a continuous random variable will take a value within any interval. A continuous probability distribution is shown as a smooth curve. The area underneath the curve is always equal to 100 percent.

The coin toss histograms approximate the normal probability distribution. The important features of the normal curve are shown in the coin toss histograms.

1. The normal probability histogram is symmetric. The highest point of the curve is the mean. The part of the curve to the right of the mean or expected value is a mirror image of the part to the left.
2. As with all continuous probability histograms, the total area underneath the curve is equal to 100 percent.
3. The curve appears to hit the *x*-axis but it never does. The chance of events very far above and below the mean or expected value is, however, very small.

3:6:3 Computing Normal Probabilities

Probabilities for a normal random variable are represented by areas underneath the normal curve. In order to compute probabilities, the normal model requires that you convert the value of the random variable for which you wish to compute a probability into standard units or *z*-scores.

A ***z*-score** tells you how many standard deviations the value of the random variable is above or below the mean.

Z-scores for values of the random variable above the mean are positive; z-scores for values below the mean are negative. For example, a z-score of −1 means the value of the random variable is one standard deviation below the mean.

Try these examples. Assume the mean life of all computer chips produced in the last year is 2,000 hours. Further assume that we know that the standard deviation is 200 hours.

Convert the value of 1,800 hours into standard units or a z-score. That is, how many standard deviations below the mean is the value of 1,800 hours?

One standard deviation is 200 hours and 1,800 hours is 200 hours less than the mean. Thus 1,800 hours converts to −1 standard units or a z-score of −1.

Convert the value of 2,300 hours into standard units or a z-score. That is, how many standard deviations above the mean is the value of 2,300 hours?

Twenty-three hundred hours is 300 hours above the mean which is 1.5 standard deviations above the mean. The raw value of 2,300 hours is equivalent to +1.5 standard units.

Standard units or z-scores tell us how many standard deviations the value of the random variable is above or below the mean. You can use the following expression to compute z-scores.

$$z\text{-score} = (\text{value of random variable} - \text{mean}) \div \text{standard deviation}$$

A standard normal curve is shown below.

Figure 3-10 A Standard Normal Curve

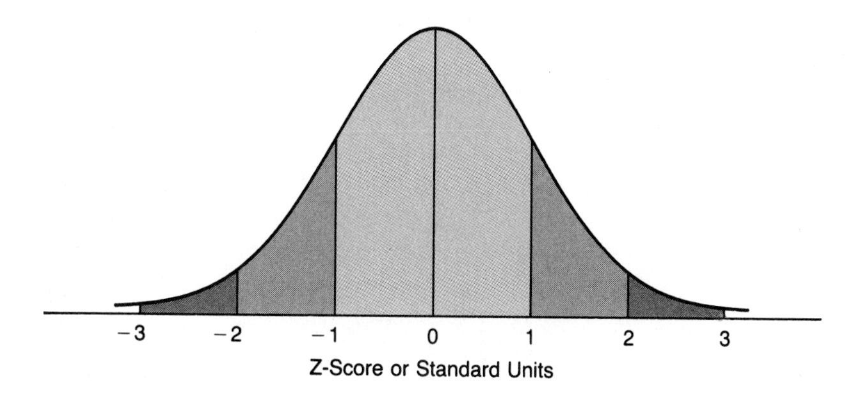

The area under the normal curve between −1 and +1 standard units is 68.26 percent. The area under the normal curve between −2 and +2 standard units is 95.44 percent. The area under the normal curve between −3 and +3 standard units is 99.72 percent.

Why isn't the area between plus or minus two standard units twice as great as between plus or minus one standard unit?

The bell-shaped curve tells us that events that are further away from the mean are less likely. Suppose you can represent, or model, the height of all males by the normal curve. If the average height is 70 inches wouldn't you expect more males between 67 and 70 inches than between 60 and 63 inches?

In the back of most statistics books you will find an "area under the normal curve" table. A small portion is shown on page 110.

Area Under Normal Curve—A Small Table

Number of Standard Units (z-scores)	Area Between the Mean and the Number of Standard Units Away from the Mean
.5	19.15 percent
.75	27.34 percent
1.0	34.13 percent
1.2	38.49 percent
1.5	43.32 percent
1.6	44.52 percent
1.75	45.99 percent
2.0	47.72 percent
2.5	49.38 percent
2.8	49.74 percent

Since the normal curve is symmetric, the areas are the same for positive and negative z-scores. You're now ready to calculate some normal probabilities.

Problem 1. Return to the computer chip example. We are interested in computing probabilities for the continuous random variable, the life of a computer chip. It is continuous because it can take on a very large number of possible values. If you select a computer chip at random, what are the chances of getting one that will last between 1,800 and 2,200 hours? *Assume* that the mean and standard deviation are 2,000 hours and 200 hours, respectively.

You should use the following approach when computing normal probabilities:

1. Convert the values of the random variable to z-scores or standard units.
2. Draw a sketch of the normal curve and shade in the desired area.
3. Use the table of areas under the normal curve to compute the exact probability.

You can use your sketch to help you identify the area under the curve that represents the desired probability. If you sketch it carefully it also serves as a good estimate to check your computation.

Returning to Problem 1, you first convert 1,800 and 2,200 hours into z-scores. Eighteen hundred hours is equivalent to a -1 z-score and 2,200 hours is equivalent to a $+1$ z-score. Here's the sketch of the normal curve.

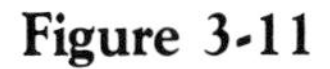

Figure 3-11

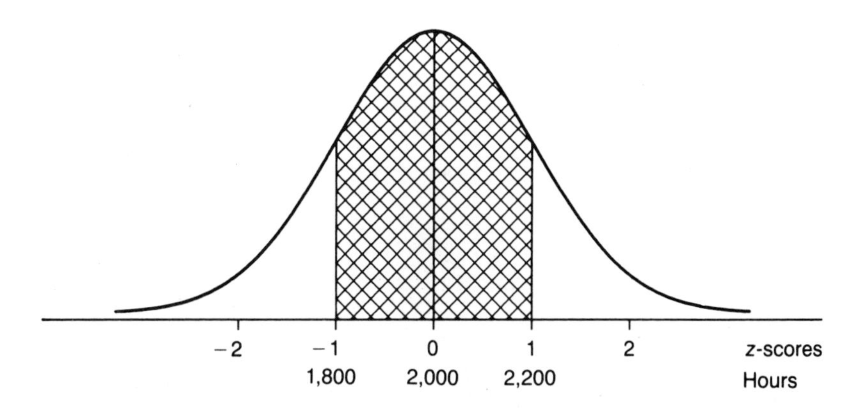

The cross-hatched area is between plus and minus one standard units. The area is 34.13 percent + 34.13 percent or 68.26 percent and the probability is .6826.

Problem 2. What are the chances of randomly selecting a chip that lasts more than 2,300 hours? Please solve this problem in the space below. Remember to convert 2,300 hours to a z-score, draw a curve and shade in the desired area, and then use the area under the normal curve table.

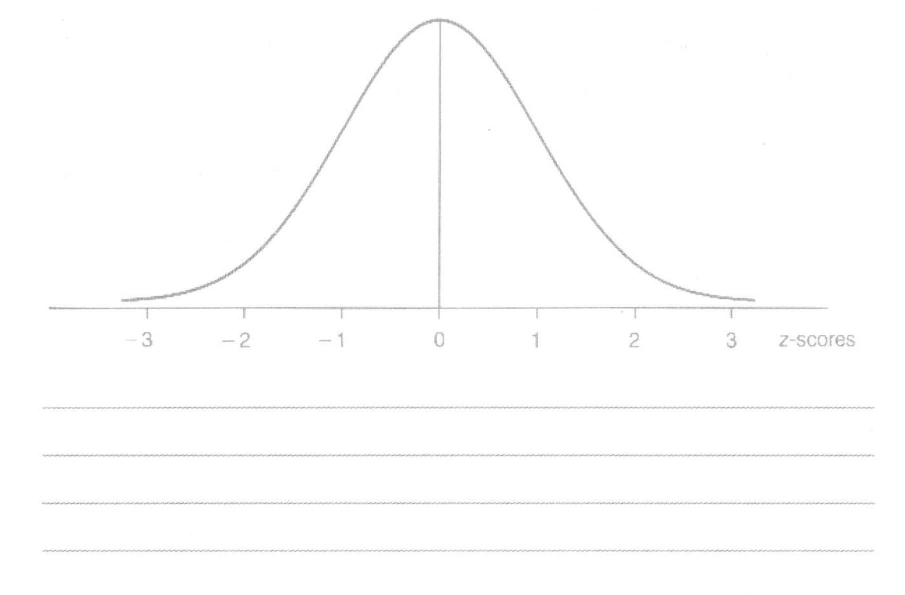

Figure 3-12

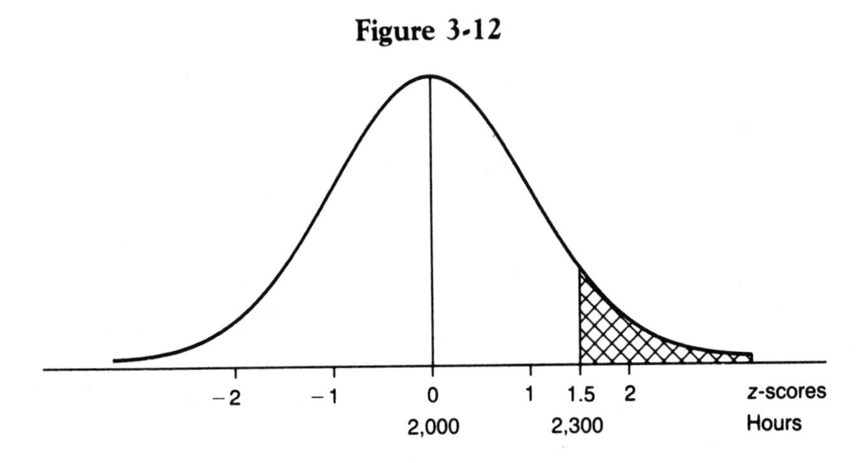

The cross-hatched area is to the right of +1.5 standard units. We know from our shortened normal table that the area between zero and +1.5 standard units is 43.32 percent. Since the area of each tail of the curve is 50 percent, we can determine the probability as follows.

$$50 \text{ percent} - 43.32 \text{ percent} = 6.68 \text{ percent or } .0668$$

Problem 3. What are the chances of randomly selecting a chip which lasts between 1,600 and 2,100 hours? Please solve this in the space below.

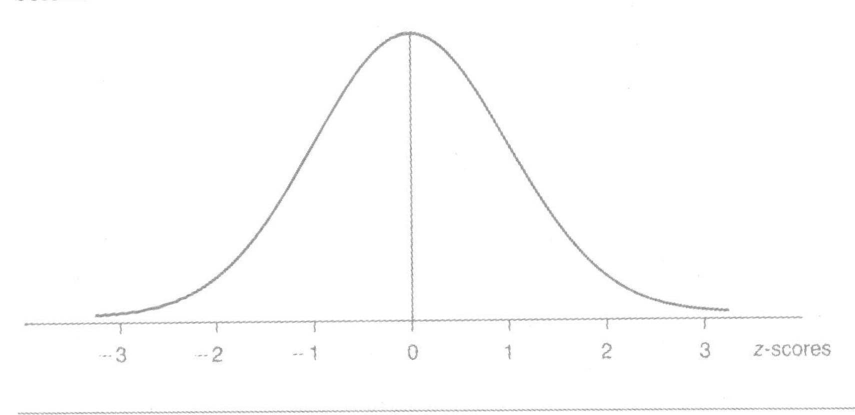

Figure 3-13

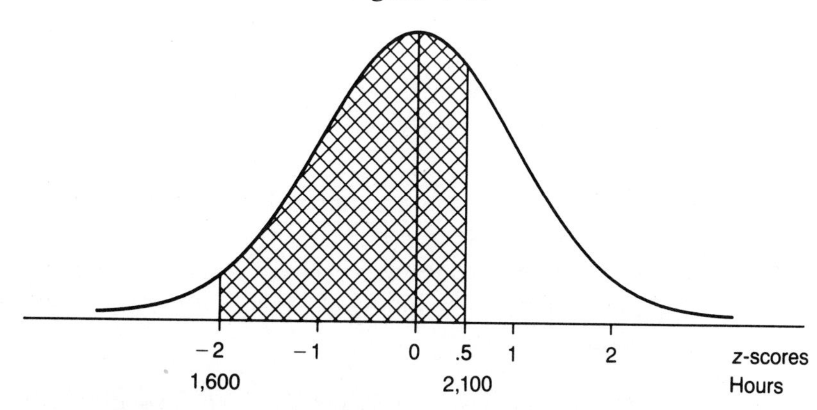

The cross-hatched area consists of two parts. One is between zero and −2 standard units. The other is between zero and +.5 standard units. The desired probability is the sum of these two areas.

47.72 percent + 19.15 percent = 66.87 percent or .6687

Problem 4. What are the chances of randomly selecting a chip which lasts exactly 1,900 hours? Sketch this out and determine the probability before reading on.

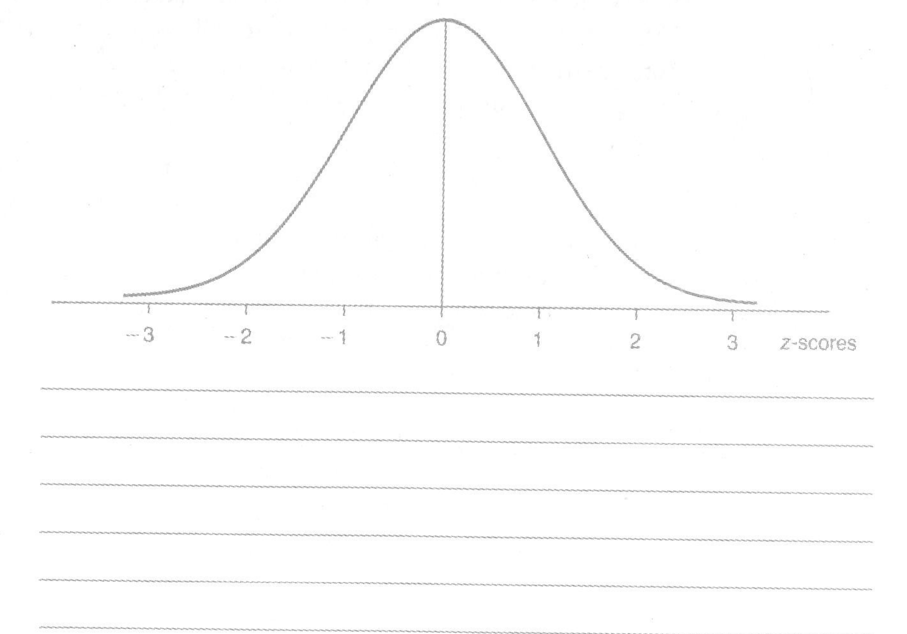

How much area is underneath the normal curve at 1,900 hours? I hope you concluded none. The area under a point is zero. We can generalize and say:

> the probability that a *continuous* random variable will take on exactly a single value is zero.

3:6:4 Decision-Making and the Normal Curve

I first mentioned using probability models to help make decisions in the section on the binomial model. Now you'll see how the normal model can be used.

Which Express Service Should You Use? Two overnight express services are in head-to-head competition. Federal Service advertises that its average delivery time is 30 hours, while United Express claims that its average delivery time is 34 hours. However, United Express arrival times are more consistent in that its standard deviation is only two hours whereas Federal's standard deviation in delivery time is five hours. Which service should you use if you want your document to arrive within 36 hours?

The amount of time it takes for a document to arrive can probably be modeled by a normal curve. There are many factors (weather, traffic, courier breakdowns, etc.) that will cause arrival times to vary from one delivery to the next. If you kept records and plotted a histogram of arrival times, it would look very similar to a normal curve.

You want to select the service which has the highest chance of meeting the 36 hour deadline. You need to compute the chances of your document arriving within 36 hours by both services and select the service with the highest probability.

Take some time and try to draw the normal curves, shade in the appropriate areas, and solve the problem. Please don't look ahead.

Here's the solution.

Below are the two normal curve sketches. The area under each curve is 100 percent.

Figure 3-14 Probability Distributions for Two Delivery Services

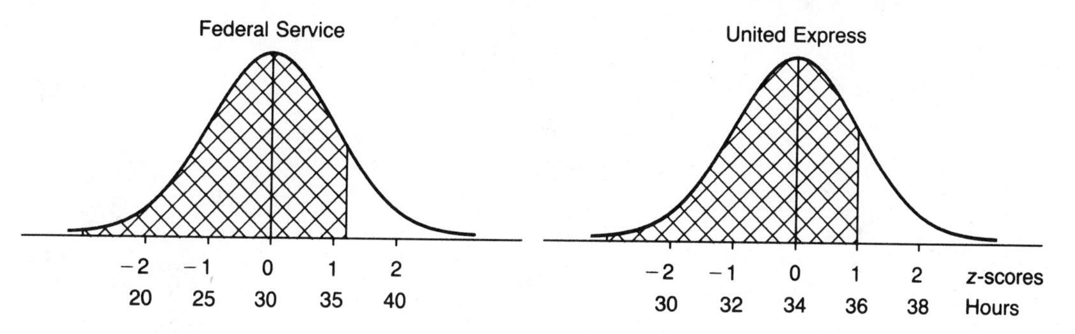

If your sketches are reasonably accurate, you should be able to see that the cross-hatched area for Federal Service is slightly larger. This means they have a higher likelihood of getting your document to its destination within 36 hours. The area to the left of 36 hours in each curve consists of two segments. The lower tail of the curve has 50 percent of the area. The other part is between the mean and 36 hours.

For United Express, 36 hours is equal to +1 standard unit. The area between zero and one standard units is 34.13 percent. The probability of delivery within 36 hours is

$$50 \text{ percent} + 34.13 \text{ percent} = 84.13 \text{ percent or } .8413.$$

For Federal Service, 36 hours is equal to +1.2 standard units. The area between zero and 1.2 standard units is 38.49 percent. The probability of delivery within 36 hours is

$$50 \text{ percent} + 38.49 \text{ percent} = 88.49 \text{ percent or } .8849.$$

Go with Federal Service.

Should you still select Federal Service if you could wait up to 38 hours? Again sketch the normal curves, cross-hatch the appropriate areas, and find the probabilities.

The probability that United Express will deliver the document within 38 hours is .9772. Federal Service's chances are .9452. In this case, you should select United Express. It all depends on how much time you have.

In Search of New Species. In 1912, Charles Dawson, an amateur archeologist, turned up portions of a skull which appeared to have the cranium of a man and a lower jaw of an ape. He argued that this creature was the missing link between man and ape, and he became widely acclaimed. As the years passed, the Dawson fossils became suspect. In 1953 three British scientists concluded that the "missing link" had been a hoax. Dawson had attached the head of a man to the jaw of an ape. He had filed the teeth to disguise them.

How do reputable archeologists work? In searching for a new species, an important characteristic is brain size. Why is brain size so important in classifying species? A small brain cannot hold as many brain cells as a large one. Less obvious is the fact that larger brains can have more complex linkages between the cells which is a measure of the quality of the brain.

Here are some hypothetical (but reasonably accurate) data for a species called Australopithecus who walked on Earth about two million years ago. Their average brain size was 508 cubic centimeters (cc) and the standard deviation in brain size was about 25 cc. For your information, the brain size of modern man is 1,200 to 1,500 cc. Most of us have about 1,500 cc before Statistics and about 1,200 cc after Statistics.

Suppose a skull is found that has the same shape as Australopithecus but is 578 cc. Should the archeologist conclude that this skull belonged to the Australopithecus species or should he or she conclude that it does not? Of course, an archeologist would use additional information (such as tooth and jaw dimensions) to arrive at a decision, but for the sake of simplicity, let's assume there are no other data.

Suppose we plot the brain size data based upon previous finds and conclude that the brain size of Australopithecus can be represented by a normal curve. What are the chances that a 578 cc brain belongs to the Australopithecus species if the average size is 508 cc with a standard deviation of 25 cc? In order to calculate a probability we must rephrase the question. What are the chances of Australopithecus having a brain size of 578 cc *or more*? "Or more" is added because we know the chances of having a brain size of *exactly* 578 cc is zero. Remember, the chance of a *continuous* random variable taking on a single value is always zero. Below is the normal curve.

Figure 3-15 Distribution of Brain Size of Australopithecus

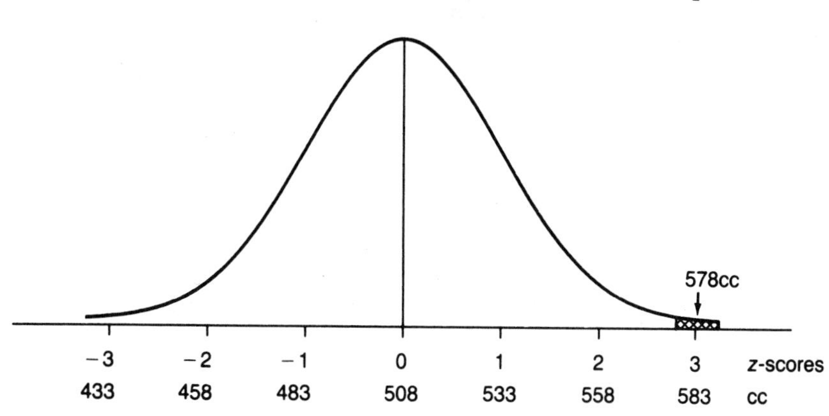

The value of 578 cc is equivalent to +2.8 standard units. The area to the left of +2.8 units is equal to 50 percent + 49.74 percent = 99.74 percent. Thus the area above 578 cc is 100 percent − 99.74 = .26 percent or .0026. The probability of getting a brain that big or bigger if it came from the Australopithecus species is very small—only 26 in 10,000. We just might have discovered a new species.

3:6:5 So How Do We Really Know the Mean and Standard Deviation?

In order to use the normal tables you must convert the random variable's values into standard units or *z*-scores. But you must know the mean and standard deviation of the normal curve to make the conversion. How would you know what the mean and standard deviation are?

Before answering that question, return to the computer chip example. Picture all the computer chips that the company has in stock or in production. Can you see them stored in the warehouse awaiting shipment to computer manufacturers and retailers? Can you see them being produced in sterile "white rooms?" These constitute a population of computer chips.

A **population** is the total set of objects about which information is wanted.

The information that we want is the mean and standard deviation. We assumed that the mean and standard deviation were 2,000 and 200 hours respectively. How could you determine this? You could life-test every computer chip until it failed and record the number of operational hours, then compute the descriptive statistics—the mean and the standard deviation. There are some practical problems with this strategy. What are they?

First, it would take a very long time to life-test each computer chip. Even if you did, you wouldn't have any chips to sell to computer manufacturers and retailers. Can you imagine the following as your advertisement?

> Outtel Corporation is extremely proud to announce its new super-improved chip. We guarantee that it will last, on the average, 2,000 hours. That is better than 300 hours longer than its major competitor. Unfortunately, we don't have any to sell because we tested all the chips to determine the average life. You really would have liked this chip. Too bad.

The only way you can know for sure the mean and standard deviation of a population is to test the entire population. With a very few exceptions, this is not practical.

How could United Express and Federal Service have really known the average and standard deviation for their delivery times? They would have had to keep records of *each* delivery ever made. Since they are in the delivery business and not the statistics business, it is unlikely that they would have such records.

Generally speaking, we never know what the mean and standard deviation of a population are. We can, however, estimate them using the method of formal statistical reasoning. That will be our next topic.

Please review the objectives for this unit. If you have not mastered them all, review the material and then complete the exercise set below. With this chapter we finish probability; the next topic is statistics.

Exercise Set for 3:6

1. A bottling company produces 16 ounce bottles of a soft drink. Assume that the mean (or long-run average) is 16 ounces and the variance is .09. What are the chances of randomly selecting a bottle with less than or equal to 15.7 ounces of soft drink? Convert to standard units, draw a sketch, estimate the probability, and compute the exact probability.

   Note that you are given the variance, not the standard deviation. You will need to compute the standard deviation from the variance. This type of problem is called in the trade a "technical trick question." Now that you have seen it, you should never be fooled again.

2. Referring back to Problem 1, what are the chances of selecting a bottle between 15.85 and 16.60 ounces? Convert to standard units, draw a sketch, and compute the exact probability.

3. Still dealing with Problem 1, what are the chances of randomly selecting a bottle with exactly 16.09876859402345 ounces?

4. You have been asked by your mother to mop the floor. You are not thrilled about this opportunity. She tells you that you can use one of two approaches—the sponge mop or the "knees to the floor." She has kept meticulous records:

sponge mop	average time: 10 minutes	standard deviation: 2.5 minutes
knees/floor	average time: 14 minutes	standard deviation: 1 minute

Your favorite soap opera is going on in 15 minutes. Which approach should you use if you want to maximize the chances that you will be finished before the soap opera goes on? (Assume we have ruled out getting your sister or brother to do it.) Draw two sketches, estimate the probabilities, compute the exact probabilities, and make your decision.

4.

Sampling, Confidence Intervals, and Hypothesis Testing

4:1 INTRODUCTION

In the last chapter you computed normal probabilities. You *assumed* that you knew the mean and standard deviation. Sometimes this will be the case, but often it will not. After all, how would you know the mean lifetime of a computer chip without life testing each and every chip? How would you know the actual percentage of voters who will vote for a candidate unless you polled every eligible voter in the United States?

What do you do when you don't know the mean and standard deviation of a normal distribution? You *estimate* them using sampling procedures and the methods of inductive inference shown in this chapter.

4:2 SAMPLING AND WHAT CAN GO WRONG

By the end of this unit you should be able to:

1. Distinguish, in your own words, between a population and a sample.
2. Distinguish, in your own words, between a population parameter and a sample statistic.
3. Differentiate between inductive and deductive inferences.
4. Explain to others why a sample statistic might not be close to the population parameter it is estimating.
5. Overcome some of the more common biases in sampling.
6. Distinguish, in your own words, between population standard deviation and sampling bias.

Here are two examples that illustrate the ideas behind sampling:

1. A computer company wishes to estimate the average operational lifetime of a shipment of computer chips it has just received. This is the first batch from a new supplier and the company is concerned that the chips may not last as long as those from their previous supplier.
2. You want to know who will win an upcoming election. You ask a reputable polling organization to conduct an opinion survey.

Let's begin with some very important definitions:

A **population** is the entire group of objects about which you want information.

A **sample** is a part of the population. It is used to gain knowledge about the population.

A **population parameter** is a numerical fact about a population. It is a fixed or constant value and you almost never know its actual value.

A **sample statistic** is a numerical fact about a sample. It is used to estimate the unknown population parameter. Even though the population parameter is constant, the sample statistic can vary from sample to sample.

What is the population in the computer chip example? It is all the future shipments of computer chips from the new supplier. What is the population parameter you are interested in knowing? It is the average lifetime for all chips. It is a fixed value and we don't know what that value is. The first shipment is a sample or a part of the population. If you can assume that the first shipment is representative of all future shipments, you can compute the mean of the sample and use it to estimate the average lifetime of all future chips. The sample mean is the sample statistic used to estimate the population parameter. Remember, sample statistics are what we know and population parameters are what we want to estimate.

Go through the same exercise for the second example. What are the population, population parameter, and sample statistic? Write your answers in the space provided. Please don't look ahead.

Hopefully, you have the following answers:

1. Population: U.S. residents 18 years and older who are registered voters.
2. Population parameter: The actual percentage who will vote for each candidate on election day.
3. Sample statistic: The percentage of voters in the sample who said they would vote for each candidate.

The sample mean and sample percentage are two common sample statistics. While similar, there are two differences. First, sample means can take on all values. They can be negative, such as the average temperature in January at the North Pole, or positive, such as the average lifetime of a computer chip. Sample percentages are numbers between zero and one. Secondly, you compute sample means when you have quantitative data such as the number of hours a computer chip lasts. You compute sample percentages when you have qualitative data such as "will you vote for the Republican candidate—yes or no?"

In Chapter 3 we made **deductive inferences**. You make deductive inferences when you know the population parameter(s) and you compute specific probabilities. You are using general knowledge about the population—its parameters—to make valid, specific probability statements.

When you don't know the value of the population parameter and wish to estimate it based upon a small sample from the population, then you are making **inductive inferences**. Here you use the specific and known information of the sample statistic to draw general conclusions about the unknown population parameter. You will be drawing inductive inferences for the rest of this book—that is what statistics is all about.

Making inductive inferences is not just something statisticians do. We all do it everyday. Say a friend offers to "fix you up" with a blind date. Your friend tells you that your date is a wonderful person with a pleasing personality. But you believe that all blind dates are beasts. How did you arrive at that general conclusion?

Let's suppose that you have had a few blind dates in the past. Think of these as a sample from the population of possible blind dates. Your previous blind dates have been beasts. Think of this as your sample statistic. Based upon only a *few* previous blind dates you jump to the general conclusion that *all* future blind dates who are described as having a "pleasing personality" are beasts. You have used a statistic based upon sample evidence to draw a general conclusion about the population of blind dates. You have just made an inductive inference. If you accept the offer and your date is wonderful, you have also learned your first lesson about inductive inferences—*you can be wrong.*

Even turkeys draw inductive inferences. Suppose that in late September a turkey hears the noon whistle and then is fed. This goes on for about two months—the noon whistle and his feeding. After two months of sample data, the turkey reasons that when he hears the whistle he is fed. This is a statistic because it is based upon a sample of 60 days.

This turkey draws an inductive inference. He concludes that from now on he *will always* be fed when he hears the noon whistle. Based upon two months of data he has just drawn a general conclusion about the population of all future days. On Thanksgiving Day, the turkey also learns the lesson about drawing incorrect inductive inferences.

In making inductive inferences you estimate population parameters based upon sample statistics. How close will the estimates be? That depends upon sampling bias and the population standard deviation.

4:2:1 Bias

Bias causes the sample statistic to consistently underestimate or overestimate the population parameter. Sampling bias occurs when a mistake is made in collecting the data.

There are many types of sampling biases, but only three will be discussed here.

Selection bias. A classic selection bias blunder was made in the summer of 1936. President Roosevelt was completing his first term in office. The Republican candidate was Governor Alf Landon of Kansas. While most of the polls were predicting that Roosevelt would win, the *Literary Digest*'s poll predicted that Landon would win big. Their poll was based upon 2.4 million voters as compared to the usual sample size of 1,400 to 1,800.

Of course, Roosevelt drubbed Landon. In fact, Landon won only two states—Maine and Vermont. Up to that time, the conventional wisdom was, "As Maine goes, so goes the nation." After the 1936 election a political writer rewrote the adage to read, "As Maine goes, so goes Vermont." The *Digest*'s poll had predicted Landon to win 57 percent to 43 percent. Actually Roosevelt received 62 percent of the popular vote. If we look at how the *Digest* conducted their survey, you'll see what went wrong.

A sampling procedure should be fair, unbiased, and should select a representative sample from the population. The *Digest* sent out ten million questionnaires. They obtained the names and addresses for their sample from telephone books and club membership lists. In 1936, the Great Depression was at its worst. The families that had telephones or belonged to clubs were the well-to-do who tended to vote Republican. Thus, the *Digest* pollsters had primarily surveyed well-to-do Republicans and they were going to vote for Landon. Unfortunately for Landon, Democrats outnumbered Republicans and Roosevelt won the election. Because of selection bias, the *Digest*'s pollsters made an incorrect prediction.

A systematic tendency to exclude one kind of member of the population from the sample is called **selection bias**.

To overcome selection bias you must ensure that each member of the population has a chance of being selected. In simple random sampling each member has an *equal* chance of being selected. In a probability sample, each member has a chance of being in the final sample.

Increasing the sample size will not remove the selection bias. Actually, it only makes it worse. You can, however, reduce selection bias by using probability or simple random sampling.

Response bias. This bias generally occurs in public opinion surveys. You can change the sample statistic—the percent in favor of an issue—by merely changing the wording of the question. Roll and Cantril, in their book entitled *Polls: Their Use and Misuse in Politics*, illustrated this bias. Some citizens were asked if they favored "adding to the Constitution" a one-term limit for the presidency. Others were asked if they favored "changing the Constitution" to include a one-term limit. Fifty percent were in favor when the phrase "adding to" was used and 65 percent were in favor when the term "changing" was used.

We also know that the order in which the candidates' names appear in a survey questionnaire can affect an individual's response. The name which is placed first can often pick up five percentage points. Finally, the race of the interviewer and respondent can make a difference. A white interviewer of a black respondent may obtain a different response than a black interviewer of a black respondent. The same is true of a black interviewer and a white respondent. This is especially true when the questions deal with black-white relations.

It is important to remember that increasing the sample size will not remove the response bias. Actually it only makes it worse. Probability sampling will not help, for the problem is not how the sample was selected, but how the questions are worded or presented and who asks them.

Nonresponse bias. In the *Digest* survey of ten million voters, only 2.4 million (or 24 percent) responded. A large number of nonrespondents can create a serious bias in the sample statistic. The nonrespondents often differ from those that responded beyond the fact that they didn't respond. What might account for the 7.6 million voters who did not respond to the *Digest* survey? They might have been poorer and couldn't afford the postage to return the questionnaire; they might have been illiterate and couldn't read the questionnaire; or they might have been richer and away on a family vacation when the questionnaire arrived. In any case, the final sample of 2.4 million respondents might not even be representative of the ten million individuals who were mailed the questionnaire.

Studies have shown that both lower and upper income level families tend not to respond to mailed surveys. This means that middle income families are over-represented in the final sample—an example of nonresponse bias. Nonresponse bias even occurs in face-to-face interviews. Those people who are not at home when the interviewer stops by may be systematically different from those that are home. Good survey organizations, such as the Gallup organization, have developed ways to deal with this problem.

The nonresponse bias is different from selection bias. The ten million voters who were mailed the *Digest* questionnaire are called the **sampling frame**. In selection bias we are concerned that the sampling frame may not be representative of the population. In nonresponse bias, we are concerned that the respondents may not even be representative of the sampling frame. It is important to know that increasing the sample size by resampling those who did not initially respond will reduce the nonresponse bias. At a minimum, you should determine if respondents and nonrespondents differ systematically in terms of age, race, or socioeconomic status. If they do, you should probably not use your data to make inferences.

4:2:2 Population Standard Deviation

Population standard deviation measures the spread in the values of the observations within the population. The spread is not due to sampling mistakes made during data collection. It is due to thousands of factors which cause the values within a population to vary from one another and from the population parameter.

Suppose we tested every computer chip from a shipment. Why wouldn't you expect them to last the same amount of time? Think about it and please write your answer below.

Hopefully, you said because it is impossible to hold everything constant when you produced the chips. For example, different assemblers worked on the chips at different times of the day. During the production run, some of the assemblers may have gotten tired. The raw material may have varied slightly from chip to chip. There are a thousand small differences from one chip to the next. Each of these factors may have a small impact on the life of a chip. That is what accounts for population standard deviation. If we could eliminate all sources of variability the population standard deviation would be zero. In practice, this is impossible.

Suppose we took a sample of 25 computer chips at random from the population. Because of population standard deviation we would not expect the sample mean (sample statistic) to be the same as the population mean (population parameter). Thus, even when we have no sampling biases there is no guarantee that the sample statistic will be close to the population parameter.

4:2:3 How Gallup Selects Its Sample

Most professional polling organizations do not use simple random sampling in voter preference surveys. It is not practical. Simple random sampling requires that we have a list of all the registered voters in the United States. No such list exists. Even if it did, the names drawn from such a list would be scattered all over the country. You might interview a single voter in Bangor, Maine; another in Macon, Georgia; and one in Rock Island, Illinois; Durango, Colorado; and Medford, Oregon. Can you see how expensive and time-consuming this process would be? Many polling organizations use *multistage cluster sampling*, which is a probability sampling procedure. Typically the country is subdivided into four regions as shown below.

Figure 4-1 Regional Divisions of the United States

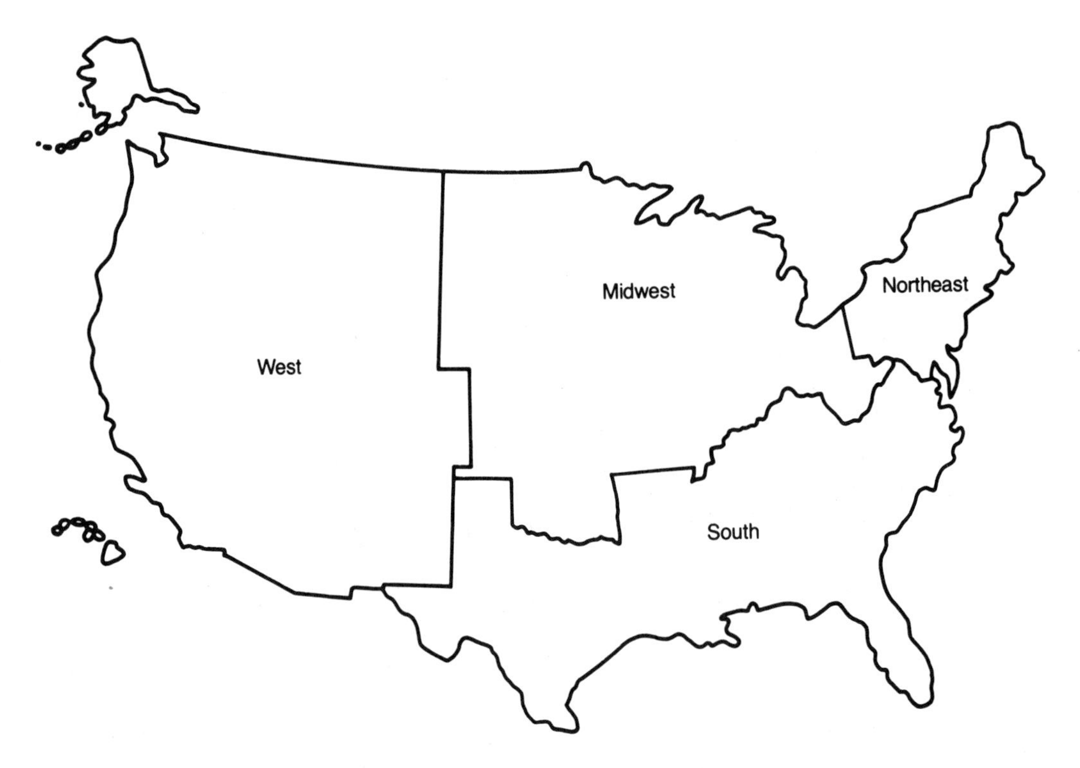

Within each region, towns and cities are clustered together by size. One cluster might be cities between 250,000 and 750,000 people. Towns are selected using simple random sampling from each cluster within a region. The cities are then subdivided into wards and a simple random sample of wards are selected. The wards are then subdivided into precincts and again a simple random sample of precincts are selected. Finally, within each ward a simple random sample of households are selected for the polling.

The term "multistage" simply means that the sampling process takes place in four stages—cities, wards, precincts, and households. The term "cluster" refers to the grouping of cities by size within each region of the country. Because multistage cluster sampling uses chance to determine which city, ward, precinct, and household will be polled, it is a probability sampling procedure. Selection bias should be minimized.

The Gallup organization only samples about 1,500 voters. Yet they have always been very accurate in their predictions—generally within two or three percentage points of the actual vote. It pays to use valid methods in selecting the sample and to word the questions properly.

Review this section and make sure that you can accomplish the objectives. Below are some exercises to help you gain mastery.

Exercise Set for 4:2

1. Below are two everyday arguments. One illustrates inductive reasoning, the other illustrates deductive reasoning. Which is which and why?

   a. This wedge of a melon is ripe. The wedge of the melon is a representative sample of the entire melon. So the entire melon is ripe.
   b. All melons that are ripe are tasty. This melon is ripe. It will be tasty.

2. Identify the population, population parameter, and the sample statistic in the following scenarios.

   a. You want to estimate the percentage of red jelly beans in a large jar. You select randomly 50 beans and find that 20 are red.
   b. You wish to estimate the average income for a family of four in a large city. You take a probability sample of 1,500 families of four and find that the average income is $20,100.
   c. A television rating service wishes to know who is watching a particular soap opera. Fifteen hundred households are asked and 100 have watched the show during the week.

3. In the 1936 *Literary Digest* polling example discussed earlier, selecting people from the telephone book led to selection bias. Would we face that problem today? Can you think of any group of people who would not be listed in the telephone book? Could this bias our sample?

4. Below are four pictures. Think of the center of the target as the unknown population parameter we are trying to estimate. The following four conditions are shown:

   high bias and low population standard deviation
   high bias and high population standard deviation
   low bias and low population standard deviation
   low bias and high population standard deviation.

Which is which and why?

Figure 4-2

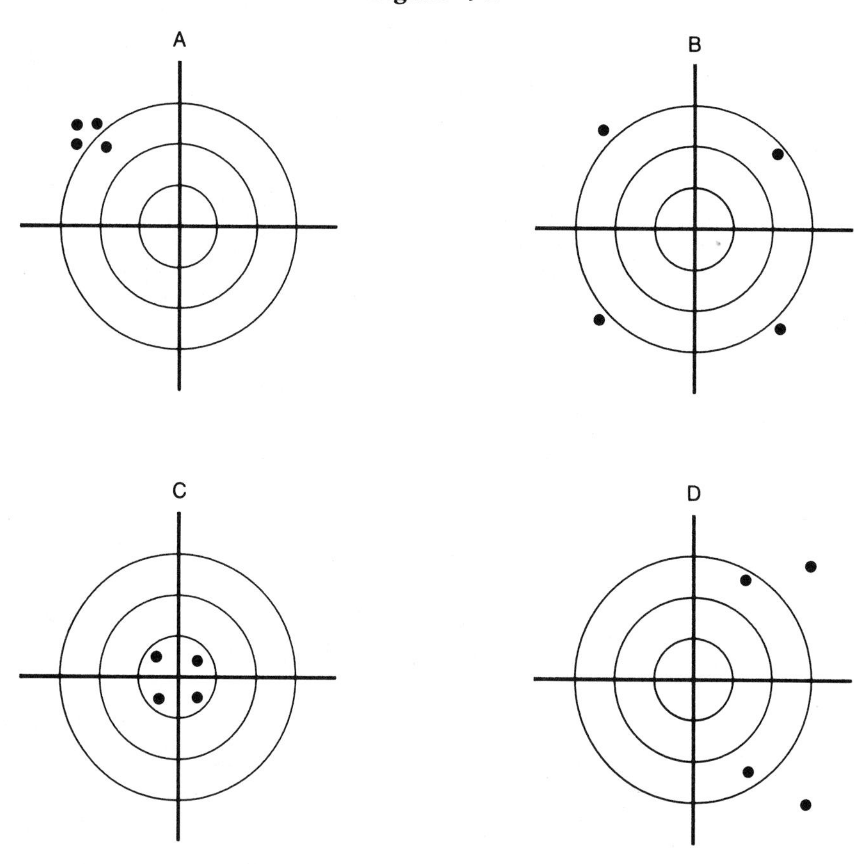

5. A firm selling small portable fans wishes to estimate how successful their new fan will be. They take a simple random sample of homeowners from the county records office and determine that 16 percent would buy their fan. Have they encountered the selection bias?

6. A company produces college rings. While the rings of a given size appear to be the same diameter, if we measured them carefully we would find that they differ slightly. What human and technological factors might account for this population standard deviation?

4:3 THE SAMPLING DISTRIBUTION AND THE CENTRAL LIMIT THEOREM

By the end of this unit you should be able to:

1. Construct a sampling distribution for the sample mean and explain what it is to others.
2. Distinguish between the mean and standard deviation of a sample, a population, and a sampling distribution.
3. Explain why the mean of the sampling distribution *must* equal the mean of the population—the population parameter.
4. Explain why the standard deviation of the sampling distribution *must* be smaller than the standard deviation of the population due to the "averaging out" effect.
5. Explain why the shape of the sampling distribution *must* be normal, provided the sample size is sufficiently large.
6. Explain how the properties of the sampling distribution are useful in estimating unknown population parameters.

Our goal in this chapter is to estimate population parameters using inductive inference methods. First, you need to comprehend sampling distributions and the central limit theorem. Let's begin with a very small population of only six workers. Here is the number of days each worker has been absent in the last quarter.

Worker A	0 days
Worker B	9 days
Worker C	6 days
Worker D	3 days
Worker E	1 day
Worker F	5 days

What are the mean and standard deviation of the population? These are always unknown when the population is large. However, for a small population we can easily calculate them. There is one minor problem. The terms "mean" and "standard deviation" can be applied to a population and a sample. So in order to differentiate the population parameter and the sample statistic, let's use the following terminology.

Terminology Table

	Mean	Variance	Standard Deviation
Sample Statistic	*x*-bar	s^2	s
Population Parameter	mu	sigma^2	sigma

Below are the calculations for the mean and the standard deviation of the population.

$$\text{mu} = (0 + 9 + 6 + 3 + 1 + 5) \div 6 = 4$$

$$\text{sigma}^2 = [(0 - 4)^2 + (9 - 4)^2 + (6 - 4)^2 + (3 - 4)^2 + (1 - 4)^2 + (5 - 4)^2] \div 6$$

$$= 9.33$$

$$\text{sigma} = \sqrt{9.33} = 3.06$$

You're probably wondering why the denominator of sigma^2 is six rather than five or $(n - 1)$. In 1:4:2 you learned that when you compute the variance or standard deviation of a *sample* the denominator is always $(n - 1)$. This is done to obtain a more accurate estimate of the unknown population standard deviation. However, in this case you are computing the standard deviation of a population directly because the population is very small. This is the only instance when the denominator is not $(n - 1)$.

Now let's take *all possible* samples of size three from the population. How many different samples of size three are there? Please use the expression for combinations from Chapter 3.

The number of possible samples of three is

$$\frac{6!}{3! \times 3!} = 20$$

Next we'll compute the sample averages for the twenty samples and draw a histogram. The frequency histogram for the sample averages is called the **sampling distribution of the sample mean.** It is a sampling distribution because it is obtained by hypothetically taking all possible samples of a specified size from the population. It is a distribution because all the sample means do not have the same value. There will be a distribution of values. Below are the 20 samples and their means. Figure 4-3 is the sampling distribution of the sample mean for samples of size three.

Sample Means for All Possible Samples with $n = 3$ Observations

Sample	*x*-bar	Sample	*x*-bar
0, 1, 3	1.33	1, 3, 5	3.00
0, 1, 5	2.00	1, 3, 6	3.33
0, 1, 6	2.33	1, 3, 9	4.33
0, 1, 9	3.33	1, 5, 6	4.00
0, 3, 5	2.67	1, 5, 9	5.00
0, 3, 6	3.00	1, 6, 9	5.33
0, 3, 9	4.00	3, 5, 6	4.67
0, 5, 6	3.67	3, 5, 9	5.67
0, 5, 9	4.67	3, 6, 9	6.00
0, 6, 9	5.00	5, 6, 9	6.67

Figure 4-3 Sampling Distribution for Samples of Size Three

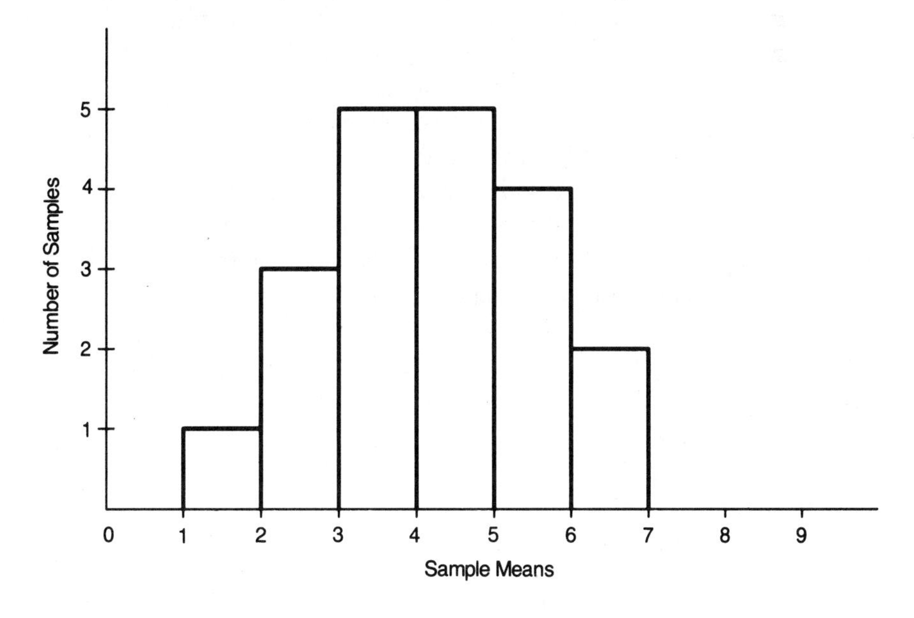

We are almost ready to determine the mean and standard deviation for the above sampling distribution of the sample means for samples of size three. But first you need some additional terminology.

Terminology Table Revisited

	Mean	Variance	Standard Deviation
Sample	*x*-bar	s^2	*s*
Population	mu	sigma^2	sigma
Sampling Distribution	mu(*x*-bar)	sigma^2(*x*-bar)	sigma(*x*-bar)

Why do we use different symbols or words for the mean or the standard deviation? If a father and son are both named Bill, how do you distinguish them? The son is called Bill Jr. Likewise, we need to distinguish between the three means.

How do the three means and standard deviations differ? If we take one sample of size three from a population, we can compute the sample average, *x*-bar, and the standard deviation of the sample, *s*. The standard deviation—*s*—measures the spread in the three observations around *x*-bar, the sample statistic.

Instead of taking just one sample of three, we took *all possible* samples of three from the population and computed the 20 sample averages. The overall average of the 20 sample averages is the mean of the sampling distribution, mu(*x*-bar). You can see that the 20 sample averages are not all the same. They vary from 1.33 to 6.67. The standard deviation of the sampling distribution, sigma(*x*-bar), measures the spread in all the sample means around mu(*x*-bar).

The mean of the six observations of the population is called mu. The spread in the six observations around mu is called sigma or population standard deviation. In the real world, the population mean and standard deviation are unknown, and our goal is to estimate them.

You are now ready to discover the properties (the mean, standard deviation, and shape) of the sampling distribution of the sample mean.

4:3:1 Mean of the Sampling Distribution

We know the mean of the population, mu, is four days. What do you think the mean of the sampling distribution, mu(*x*-bar), will be? Should it be larger, smaller, or the same as the mean of the population? Please think about it and jot down your thoughts before reading on.

Hopefully you concluded that mu(*x*-bar) should equal mu. If you take all possible samples of three from the population, the mean of the 20 sample means should be the same as the mean of the population. After all, we're using the same numbers to compute the population mean and the mean of the sampling distribution. Verify this by computing the overall mean of the 20 sample averages in the table on page 133.

You now know that:

> The mean of the sampling distribution of *x*-bar is the same as the mean of the population from which it came.
>
> mu(*x*-bar) = mu

4:3:2 Standard Deviation of the Sampling Distribution

The standard deviation of the population is 3.05 days. The standard deviation of the sampling distribution, sigma(*x*-bar), is called the **standard error of the mean**. Should the standard error of the mean be larger, smaller, or the same as the standard deviation of the population? This is a very difficult question so take your time and write your answer below. Before answering, compare the spread among the 20 sample means to the spread among the six observations in the population.

In the population the number of sick days ranged from a low of zero to a high of nine. Do the sample means exhibit the same spread? No, they exhibit less. The lowest sample mean is 1.33 days and the highest sample mean is 6.67 days. The sample means appear to cluster around the overall mean of four days. You can conclude that the standard error of the mean is smaller than the standard deviation of the population.

Now take all possible samples of size four from the population of six workers. Using the expression for combinations, there are 15 different samples of size four. In the space provided below, compute the 15 sample averages.

You already know that the mean of the sampling distribution for samples of size four should still equal four. (Check it to be sure.) Will the standard error of the mean be the same as when we took all possible samples of size three? Or will it be larger or smaller? Think about it and write your answer below.

If you computed the 15 sample means correctly, the smallest one should be 2.25 days and the largest one should be 5.75 days. Figure 4-4 shows that most of the sample means are very close to the mean of the sampling distribution. It appears that the standard error will be smaller than when we selected samples of size three. From this you can now conclude that the standard error of the mean decreases as the sample size increases.

Figure 4-4 Sampling Distribution for Samples of Size Four

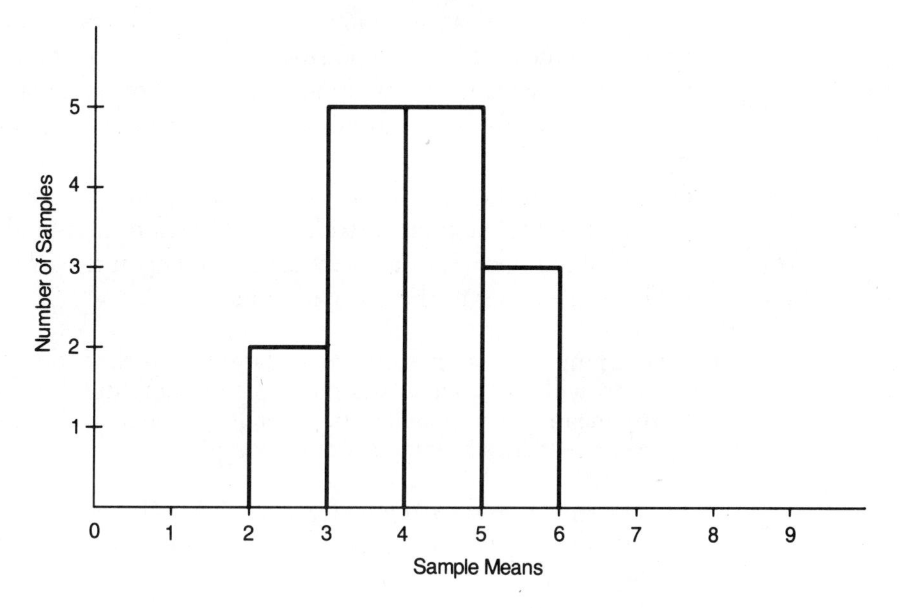

For very large-size populations the standard error of the mean is equal to the standard deviation of the population divided by the square root of the sample size.

$$\text{sigma}(x\text{-bar}) = \frac{\text{sigma}}{\sqrt{n}}$$

You can make the standard error of the mean as small as you desire by taking larger size samples.

The above expression only works for large populations. When dealing with small populations (as in our example of six), the expression is only an approximation.

Why does the standard error get smaller as the sample size increases? In order to answer that question, look at the population data once again.

Worker A	0 days
Worker B	9 days
Worker C	6 days
Worker D	3 days
Worker E	1 day
Worker F	5 days

Population values ranged from zero to nine. Many of the six numbers are not close to the population mean of four. That is why the population standard deviation was as large as 3.06 days.

The sample means for samples of size three are all close to the population mean. Is it even possible to have a sample mean of zero or nine if you take three observations from the above population? No! If you select a worker who has missed a large number of days it will be balanced out by a worker who has missed only a few days. This *averaging-out* will cause all the sample means to be relatively close to the mean of the sampling distribution. Therefore the standard error should be reduced. The sample means based upon four observations will even be closer to the population mean.

The larger the sample size, the greater the likelihood that a sample mean will be closer to the mean of the sampling distribution and the mean of the population. Therefore, the standard error of the mean decreases as sample size increases.

It's time to restate the major goal of this chapter. We want to use sample statistics (x-bar and s) to estimate the population mean (mu) and population standard deviation (sigma). How can knowing the properties of sampling distributions help you?

Sampling distributions are a theoretical concept. In practice we never take repeated samples from a population and construct a sampling distribution. It is too expensive and time consuming. For example, even for a population of only 100 elements (and that is still a very small population) there are a very large number of possible samples of three observations. Specifically, there are

$$\frac{100!}{3! \times 97!} = 161{,}700 \text{ possible samples of size three}$$

In practice we take one sample of size n from a population and estimate the unknown population mean or standard deviation. Nevertheless, sampling distributions help us.

4:3:3 What Good Is Knowing the Mean and Standard Error of the Sampling Distribution?

You wish to estimate an unknown population mean. I'll give you two choices: you can take one sample of size 3 or one sample of size 30. Cost considerations aside, which one would you take? You would probably take the larger sample. Can you relate this to the properties of sampling distributions? Hopefully, these focused questions will help.

1. If you did take all possible samples of size 3 or size 30 from the population, what would the means of the two sampling distributions equal?

They would be the same and they would both equal the unknown population mean. We know that the mean of a sampling distribution is equal to the mean of the population from which it is taken.

2. Why is the sample mean based upon 30 observations more likely to be closer to the unknown population mean? How can this be explained by the standard error of the mean?

The standard error of the mean decreases as you increase the sample size. Most of the sample means for samples of size 30 will be relatively close to the unknown population mean. Therefore, even if you take just one sample of 30 observations; it is likely to be closer to the unknown mean of the sampling distribution than one sample of size 3.

Based upon the properties of the sampling distribution, you now know that cost considerations aside, if you want to estimate an unknown population parameter, you should take the largest sample size possible.

4:3:4 The Shape of the Sampling Distribution

How does the shape of a sampling distribution compare to the shape of the population? The frequency histogram of the population and the sampling distributions for samples of size 3 and 4 are shown in Figure 4-5.

Figure 4-5 Population Frequency Histogram and Two Sampling Distributions

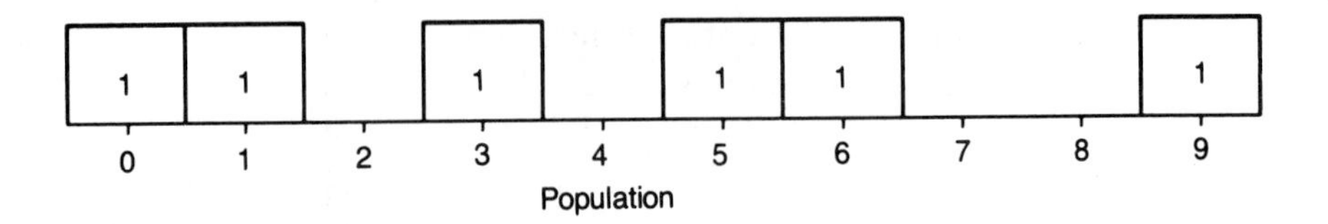

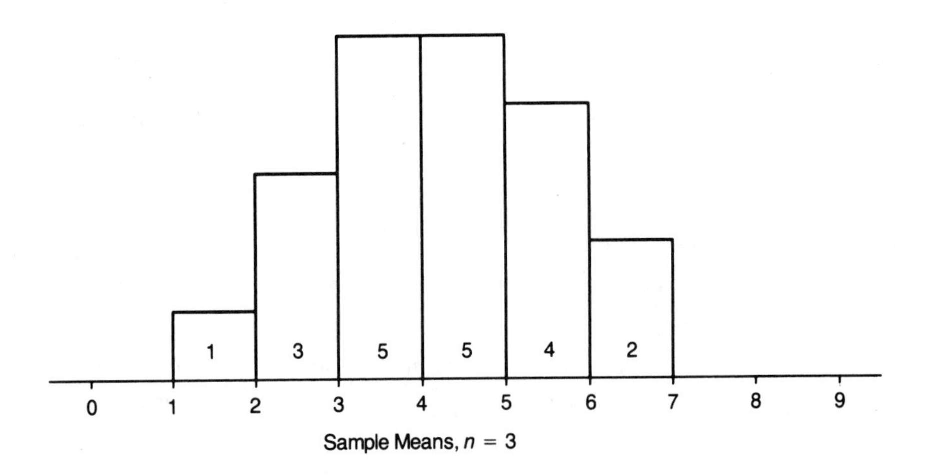

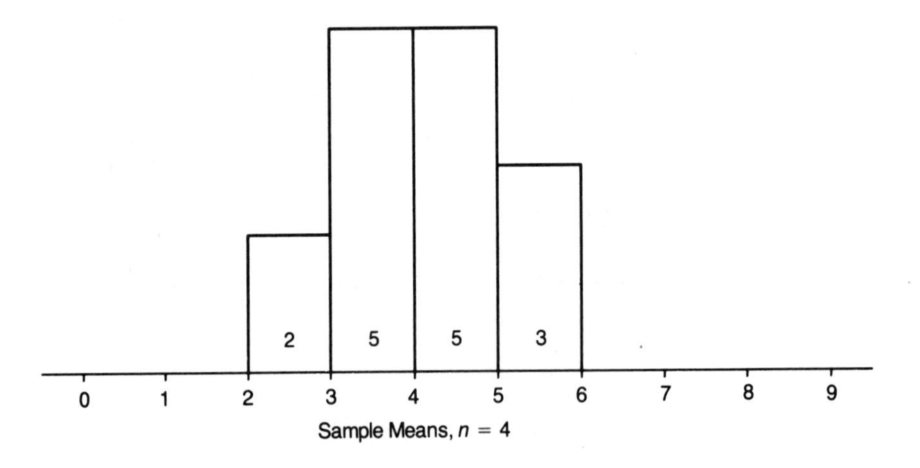

The histogram of the population is definitely not a normal curve. But notice how the shape of the sampling distribution begins to look like a normal shaped curve as the sample size increases.

This is the idea behind the **central limit theorem**. The theorem says that even if the population distribution is not normally shaped, the sampling distribution will be approximately normal. The larger the sample size, the more normal shaped the sampling distribution will be. For sample sizes larger than 30, the sampling distribution from almost any population will be normal shaped. It may be almost normal shaped for sample sizes as small as ten observations.

How does this work? As you take larger and larger samples, we expect most of the sample means to fall very close to the mean of the sampling distribution. Occasionally, a sample mean might be far away from mu(*x*-bar). But this would happen only a small percentage of the time. Sample means are as likely to fall above or below the mean of the sampling distribution. In Chapter 3, I used the same words to describe the normal curve.

The central limit theorem may not seem important, but it is. Without it, estimating population parameters from sample statistics—inductive inference—would be practically impossible.

It's time to summarize what is the most important, and perhaps most difficult, section of the book. Here are the three important lessons one more time.

1. The mean of the sampling distribution is the same as the mean of the population.
2. The standard error of the mean (or the standard deviation of the sampling distribution) for a given sample size is equal to the population standard deviation divided by the square root of the sample size. The standard error can be made as small as desired by taking larger size samples. This tells us that cost considerations aside, if you want to estimate an unknown population parameter, you should take the largest sample size possible.
3. No matter what the shape of the population distribution is, the sampling distribution will be nearly normal. For sample sizes larger than 30, the sampling distribution will be normal shaped.

4:4 DRAWING INDUCTIVE INFERENCES THROUGH CONFIDENCE INTERVALS

By the end of this unit you should be able to:

1. Explain what a confidence interval is and how it is derived from the sampling distribution.

2. Construct and interpret confidence intervals for population means and percentages.
3. Explain the need for and the impact of using Student's *t* curve in constructing confidence intervals.

You work for a tire manufacturer. The company is about to announce a new "Milemaster" tire. Your boss needs to know the average tire life for the population of Milemaster tires so that the company can set its treadwear warranty. If the warranty is set too low, the company may lose its competitive advantage. If it's too high, it may have to replace too many tires.

If the manufacturer needed to know the average tire life with absolute certainty, you could sample each and every Milemaster tire. Clearly this is not a practical approach. It takes too long, it is too costly, and there would be no tires left to sell. The only practical alternative is to take a sample from the population of tires. You can then compute the sample statistic, *x*-bar, and use it to estimate the population parameter, mu—the average life of Milemaster tires.

In this example, we'll assume that the population standard deviation is known and is 3,000 miles. This is an unrealistic assumption. If you don't know what the mean of the population is, and the standard deviation measures the dispersion around the population mean, how could you know the standard deviation? I'll relax this assumption shortly when I introduce Student's *t* curve.

Your boss asks you to explain how you will arrive at your estimate. Your plan is to take *one* simple random sample of 900 from the population of tires in the warehouse. You tell him that in a simple random sample, each tire has an equal chance of being selected. This minimizes the selection bias and ensures that the sample is representative of the population. Your boss nods his head in agreement. You select your sample of 900 tires and test them until the tread thickness is below federal standards. The average tire life for the sample, *x*-bar, is 47,500 miles. This is called a **point estimate**. Why can't we say that the unknown population mean is 47,500 miles? Please think about it and jot down your answer.

____

____

____

____

____

We know that if we took another random sample of 900 we would not get 47,500 miles again. The sample means will vary from one another due to the population standard deviation. Thus, instead of using a point estimate we will develop an *interval*, or range, estimate.

You tell your boss to imagine that instead of taking one sample of 900, you take all possible samples of 900 from the population and life test the tires. When he complains that this is impractical, you respond that you aren't really going to do this; just *imagine* it. You tell him that from the *central limit theorem*, you know that the sampling distribution will be normally distributed.

Then you tell your boss that the mean of the sampling distribution is the same as the unknown mean of the population. He is not impressed. He says that if he doesn't know the mean of the population, then he doesn't know the mean of the sampling distribution either. This time you nod your head.

Next you tell him that the standard error of the mean (the standard deviation of the sampling distribution) is much smaller than the population standard deviation due to the averaging-out effect. In fact, the standard error of the mean equals

$$\text{sigma}(x\text{-bar}) = \frac{\text{sigma}}{\sqrt{n}} = \frac{3{,}000}{\sqrt{900}} = 100 \text{ miles}$$

Now you can draw the sampling distribution.

Figure 4-6 Sampling Distribution of the Sample Mean for $n = 900$

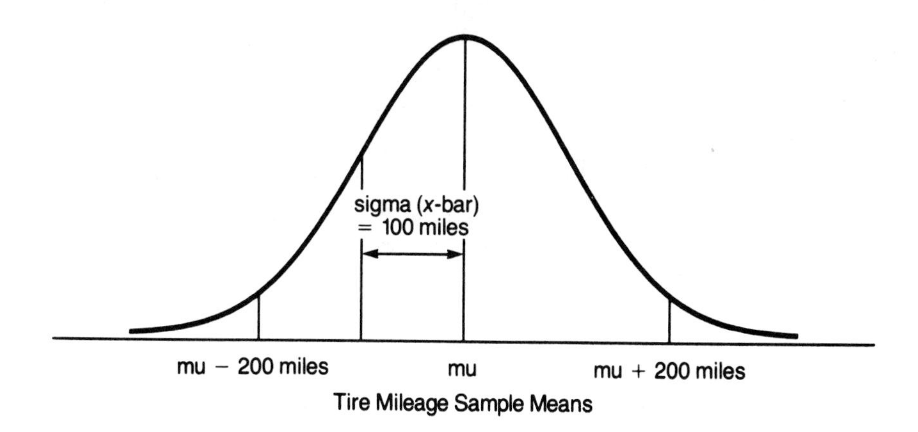

Your boss says that, while it is a lovely picture, you still don't know the mean of the population, so what good is it? You present the following arguments, which are based upon the properties of the normal curve.

1. Exactly 95.44 percent of all sample means will be within two standard errors (200 miles) of the unknown population mean.
2. The probability is .9544 that any sample mean will be within 200 miles of the unknown population mean.
3. The probability is .9544 that the sample mean we obtained from simple random sampling will be within 200 miles of the unknown population mean.
4. We are 95.44 percent confident that our sample mean of 47,500 miles will be within 200 miles of the unknown population mean.
5. We are 95.44 percent confident that the unknown population mean is somewhere between 47,300 miles and 47,700 miles (47,500 $\pm$ 200 miles).

A **confidence interval** tells us how far away the unknown population parameter can be from the sample statistic. The **confidence level** tells us how confident we can be that we are correct.

What does 95.44 percent confidence mean? If you had taken 1,000 samples of 900 and constructed 1,000 confidence intervals, you would have expected that about 954 of them would contain the unknown population mean and 46 would not. The confidence interval you actually constructed may be one of those 46 intervals. However, it is more likely that it is one of the 954 intervals that contain the unknown population mean.

Don't let it bother you that you can be wrong when you make inductive inferences. Remember the blind date! Remember the turkey! With confidence intervals you know how confident you are that you are correct. In drawing nonstatistical inductive inferences you know you may be right but you can't assign a confidence level.

We can now generalize the approach to constructing confidence intervals for the population mean.

If you take a sufficiently large simple random sample from a population (at least ten, and better yet 30 or more observations), the sampling distribution will be normal shaped. A confidence interval for mu, the population mean, is shown at the top of the page 145.

$$x\text{-bar} \pm (z \times \text{standard error of the mean})$$

or

$$x\text{-bar} \pm \left[z \times \frac{\text{sigma}}{\sqrt{n}} \right]$$

1. The term "z" is the z-score, or the number of standard deviations from the mean and is found in the normal tables. The z value depends on your desired confidence level.
2. Sigma is the standard deviation of the population which we are now assuming that we know.
3. The term "$z \times$ standard error of the mean" is called the **margin of error**. It tells you how close your sample statistic is likely to be from population parameter.

To develop additional confidence intervals you will need to use an "areas under the normal curve" table. Here's a small table you can use to construct confidence intervals in this chapter.

Table 4-1
Areas Under the Normal Curve—A Very Small Table

Number of Standard Deviations Away From the Mean (z values)	Area Between the Mean and the z Value Which Is Equal to One Half of Your Desired Level of Confidence
1.00	34.13 percent
1.038	35.00 percent
1.282	40.00 percent
1.645	45.00 percent
1.96	47.50 percent
2.0	47.72 percent
2.33	49.00 percent
2.5	49.38 percent
2.8	49.74 percent

Let's construct an 80 percent confidence interval for the average tire life of the population of Milemaster tires. An 80 percent confidence interval will have 40 percent of its area to the right and 40 percent to the left of the sample mean. The z value for an 80 percent level of confidence is 1.282 standard deviations (see Table 4-1). Therefore, an 80 percent confidence interval is equal to

47,500 miles plus or minus (1.282×100) miles.

We are 80 percent confident that the unknown population average tire life is somewhere between 47,371.80 and 47,628.20 miles. The margin of error is 128.2 miles.

4:4:1 So Who Really Knows the Population Standard Deviation?

In the tire example we assumed that we knew the population standard deviation. Now I'll eliminate that unrealistic assumption.

You can construct a confidence interval for the unknown population mean under two conditions:

1. Both the population mean and standard deviation are unknown.
2. The population mean is unknown but the population standard deviation is known.

Forget for the moment that the second condition is not realistic. Would you expect the margin of error under Condition 1 to be the same, smaller, or larger than the margin of error under Condition 2?

In Condition 2 there is only one unknown—the population mean. But in Condition 1 there are two unknowns. The more unknowns, the greater will be the margin of error. Think of the increased margin of error as the price you have to pay for not knowing the population standard deviation in addition to not knowing the population mean. Never forget, what you don't know in statistics hurts you.

The way the margin of error is increased is by using another curve—Student's t curve—to obtain the z values (except now we call them t values). For the same degree of confidence, the t values are larger than the z values. Student's t curve is a "chubby" normal curve.

Table 4-2
A Short Table of *t* Values

Sample Size Minus One ($n - 1$)	Desired Level of Confidence			
	80 Percent	90 Percent	95 Percent	99 Percent
4	1.533	2.132	2.776	4.604
9	1.383	1.833	2.262	3.250
16	1.337	1.746	2.120	2.921
25	1.316	1.708	2.060	2.787
36	1.306	1.689	2.030	2.722
Normal curve or z value	1.282	1.645	1.960	2.576

As the sample size minus one (this is called the **degrees of freedom** and I'll have more to say about it in the next chapter) gets larger, the t values get smaller. For sample sizes above 100, the t values are almost the same as the z values from the normal table. This is what Student's t curve looks like for four degrees of freedom.

Figure 4-7 Comparison of Normal and Student's *t* Curves

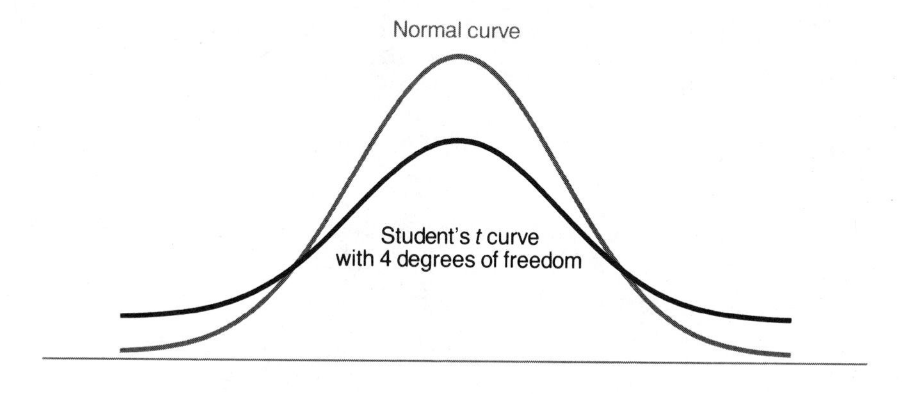

Suppose you wish to construct a 95 percent confidence interval. If you know the population standard deviation, you use the normal curve and the z value is 1.96. If you don't know the population standard deviation, you use Student's t curve. For a sample size of ten, there are nine degrees of freedom and the t value is 2.262. This will generate a larger margin of error. This is the price you pay for not knowing the population standard deviation.

If you don't know the population standard deviation, you will have to estimate it. Just as we used the sample mean to estimate the population mean, we'll use the sample standard deviation, s, to estimate the population standard deviation. The sample standard deviation is the square root of the sample variance, which is the sum of the squared differences between each of the observations and x-bar, divided by $(n-1)$. If we call the n observations $x_1, x_2, \ldots, x_n$,

$$s^2 = [(x_1 - \text{x-bar})^2 + (x_2 - \text{x-bar})^2 + \ldots + (x_n - \text{x-bar})^2] \div (n - 1)$$

$$s = \sqrt{s^2}$$

In summary, when the population standard deviation is unknown you must estimate it with the sample standard deviation and use Student's *t* curve. This will generate wider confidence intervals or larger margins of error than when the normal curve is used.

4:4:2 Constructing Confidence Intervals for the Unknown Population Mean

Let's begin with constructing confidence intervals for the mean of a population when the population standard deviation is unknown. Below is the general expression.

$$\text{x-bar} \pm (t \times \text{the estimated standard error of the mean})$$

or

$$\text{x-bar} \pm \left[t \times \frac{s}{\sqrt{n}}\right]$$

1. The value *t* is found in the Student's *t* tables (see Table 4-2). The *t* value depends on the level of confidence chosen and the number of degrees of freedom in the sample.
2. The estimated standard error of the mean is $s \div \sqrt{n}$ rather than sigma $\div \sqrt{n}$. You can't use the population standard deviation if you don't know what it is!
3. The term $t \times (s \div \sqrt{n})$ is still the margin of error.

Are ten census tracts eligible for federal aid? A census tract is a small area averaging 4,000 people. To be eligible for a federal aid program, the average income for a family of four within the ten census tracts must be less than $7,500. Only during a census year would the actual average income be known. Between the decennial census, the program administrator would have to use inductive inference procedures to determine eligibility.

A simple random sample of ten families of four is taken from the ten census tracts. In practice the sample size would never be this small. We wish to set up a 95 percent confidence interval on the population mean income for families of four in the tracts. Here's the raw data.

Incomes of Ten Randomly Selected Families

\$7,400	\$7,600
8,000	8,400
8,600	8,500
8,500	8,400
8,300	8,300

First we compute the sample mean and standard deviation for the ten observations.

$$x\text{-bar} = (7{,}400 + 7{,}600 + 8{,}000 + \ldots + 8{,}300) \div 10$$
$$= \$8{,}200$$
$$s^2 = [(7{,}400 - 8{,}200)^2 + (7{,}600 - 8{,}200)^2 + \ldots + (8{,}300 - 8{,}200)^2] \div (10 - 1)$$
$$= 164{,}444$$
$$s = \sqrt{s^2} = \$405.52$$

This is the *sample* standard deviation which is an estimate of the *population* standard deviation, sigma.

You must be able to distinguish between the standard deviation of a single sample and the estimated standard error, that is, the estimated standard deviation of the sampling distribution. We know the latter must be smaller due to the averaging-out effect discussed earlier. The sample standard deviation is \$405.52 and the estimated standard error of the mean is $405.52 \div \sqrt{10}$ or \$128.24.

Here's the 95 percent confidence interval for the mean income for families of four in the ten tracts. You can get the t value from Table 4-2.

$$x\text{-bar} \pm (t \times \text{estimated standard error})$$

$$t(\text{95 percent confidence, 9 degrees of freedom}) = 2.262$$

$$\$8{,}200 - (2.262 \times \$128.24) = \$7{,}909.92$$

$$\$8{,}200 + (2.262 \times \$128.24) = \$8{,}490.08$$

We are 95 percent confident that the unknown mean income is between $7,909.92 and $8,490.08. The margin of error is $290.08.

It's decision-making time. Are the tracts eligible for aid? Why or why not? Please write your answer below.

The ten tracts are not eligible. The reason is that we are 95 percent confident that the mean income for families of four is over the $7,500 upper limit; more specifically, between $7,909 and $8,490. Therefore the tracts are not eligible. Suppose you had obtained the following two intervals, what could you conclude?

1. 95 percent confidence interval between $6,810 and $7,390
2. 95 percent confidence interval between $7,310 and $7,890

Think it through before answering.

Since the lower and upper values of the first confidence interval are below the $7,500 maximum limit, you would conclude that the tracts are eligible. The second interval spans the $7,500 maximum. Since the population mean may fall anywhere between the lower and upper values of the confidence interval we cannot tell if the tracts are eligible; if the population mean were below $7,500 they would be eligible, otherwise they would not. If you take a larger sample you'll reduce the estimated standard error. This reduces the margin of error. Eventually it would be so small that the upper and lower values of the confidence interval would be either smaller than $7,500 (eligible) or greater than $7,500 (ineligible). Remember that the standard error of *x*-bar can be made as small as desired by increasing the sample size.

Just how fast is a new microprocessor? A computer company has a standard computer chip that can do math operations, but it is relatively slow. It takes 20,000 microseconds, or millionths of a second, to compute square roots. The company is testing a new fast chip. How fast is it? Since they don't know, they wish to estimate the mean time for computing square roots for the population of new chips. Clearly it is impractical to test each and every chip. It is inductive inference time.

We plan to take a simple random sample of 17 chips from the production line, test them, and record the execution times for computing square roots. Then we'll construct a 99 percent confidence interval on the unknown mean time for the population of new chips. The hypothetical data are shown in microseconds.

Execution Times in Microseconds

35	35	37	37
36	36	36	36
33	36	39	36
31	36	36	41
36			

Again we begin by computing the sample mean and standard deviation of the 17 observations.

$$x\text{-bar} = (35 + 35 + 37 + 37 + 36 + \ldots + 41 + 36) \div 17$$

$$= 36 \text{ microseconds}$$

$$s^2 = [(35 - 36)^2 + (35 - 36)^2 + \ldots + (36 - 36)^2] \div (17 - 1)$$

$$= 4.5$$

$$s = \sqrt{s^2} = 2.12 \text{ microseconds}$$

Remember, the standard deviation of the sample is an estimate of the population standard deviation, sigma.

We know the estimated standard error must be smaller than the estimated standard deviation of the population due to the averaging-out effect. The estimated standard error, $2.12 \div \sqrt{17}$, is .514 microseconds. We are ready to construct the 99 percent confidence interval. Remember to look up the *t* value for a 99 percent confidence interval and 16 degrees of freedom in Table 4-2.

$$x\text{-bar} \pm (t \times \text{estimated standard error})$$

$$t(\text{99 percent confidence, 16 degrees of freedom}) = 2.921$$

$$36 - (2.921 \times .514) = 34.50 \text{ microseconds}$$

$$36 + (2.921 \times .514) = 37.50 \text{ microseconds}$$

We are 99 percent confident that the unknown execution time for taking square roots for the population of new chips is between 34.50 and 37.50 microseconds. The margin of error is only 1.50 microseconds.

4:4:3 Constructing Confidence Intervals for the Unknown Population Percentage

Often we wish to draw inductive inferences about percentages or proportions. What percentage of all voters favor a certain Democratic candidate? Unlike sample means we compute sample percentages as follows.

$$\text{sample percentage} = r \div n$$

The term "*r*" is the number of people who favor the Democratic candidate. The term "*n*" is the total sample size.

Since it is impractical to sample the entire population, we take a simple random sample, compute the sample percentage or proportion, and draw an inductive inference to the unknown population proportion.

The symbol for the unknown population proportion is *p*. The symbol for the sample proportion is *p*-hat. (It seems as if it should be called *p*-bar, similar to the sample mean, *x*-bar, but it isn't.)

How do we draw inductive inferences on the population proportion? Again we use the concept of a sampling distribution. Imagine taking all possible samples of size *n* from the population of eligible voters and computing the sample proportions of voters supporting the Democratic candidate. Random sampling from a large population to see who favors a certain candidate is a Bernoulli process. The sampling distribution is a binomial distribution. But the normal model is a very close approximation of the binomial model if the sample size is large enough. A sample size of at least 100 will have a sampling distribution that is approximately normal as long as the true population proportion is between .1 and .9.

The mean of the sampling distribution is equal to the unknown population proportion you are trying to estimate. The standard error of the sample proportion is equal to

$$\sqrt{\frac{p \times (1 - p)}{n}}$$

As in the sampling distribution of *x*-bar, the standard error can be made as small as desired by increasing the sample size. However, there is one major problem in computing the standard error for the sample proportion. Look at the expression. Can you see what the problem is? Take a few minutes and think about it.

____

____

____

____

____

____

In order to compute the standard error, you must know the population proportion, *p*. But if you know the population proportion, you wouldn't need to construct a confidence interval. So in practice the standard error of *p*-hat is estimated by substituting the sample proportion for the unknown population proportion. The estimated standard error is:

$$\sqrt{\frac{p\text{-hat} \times (1 - p\text{-hat})}{n}}$$

Here's how you construct a confidence interval for the unknown population proportion.

p-hat $\pm z \times$ estimated standard error of the sample proportion

or

$$p\text{-hat} \pm z \times \sqrt{p\text{-hat} \times (1 - p\text{-hat}) \div n}$$

Who will win the election? Candidate X asks her chief-of-staff to conduct a poll. The chief-of-staff randomly selects 1,000 people and 570 indicate they will vote for Candidate X. Let's construct a 90 percent confidence interval and determine what the poll predicts. Begin by computing the sample statistic, the sample proportion. We can then use it to compute the estimated standard error.

$$p\text{-hat} = 570 \div 1{,}000 = .57$$

The estimated standard error is

$$\sqrt{(.57 \times .43) \div 1{,}000} = .0156$$

The confidence interval for the unknown population proportion is

$$p\text{-hat} \pm z \times \text{estimated standard error}$$

$$.57 \pm (1.645 \times .0156)$$

The z value of 1.645 is from the normal table, Table 4-1. The margin of error, $z \times$ the estimated standard error, is .0257. Thus the confidence interval is between

$$.57 - .0257 \text{ and } .57 + .0257.$$

We are 90 percent confident that Candidate X will receive between 54.43 percent and 59.57 percent of the vote. Please note that I changed the proportions into percentages because they are easier to comprehend.

Since you only need one vote in excess of 50 percent to win, we believe that Candidate X will win the election. This assumes that the polling was done properly. If the poll had been worded as follows, the results would be suspect. "If the vote were held today, wouldn't you want to vote for Candidate X, not her incompetent opponent, Candidate Y?" Remember, when sampling biases are present, the data are suspect and should not be used to draw inductive inferences.

We construct confidence intervals to make inductive inferences about population parameters from sample statistics. Inductive inference is what statistics is all about. If you haven't mastered the objectives at the beginning of this unit, please reread the material and then do the exercise set below.

Exercise Set for 4:4

1. Why is it necessary to resort to confidence intervals to estimate unknown population parameters?

2. You have your choice of taking one sample of size 30 or one sample of size four. What is one important property of the sampling distribution that should cause you to take the larger size sample if you wish to estimate the unknown population mean?

3. Under what conditions, if any, can the standard error be larger than the population standard deviation? Defend.

4. You wish to estimate the mean grade point average at a university. You select a simple random sample of ten students and obtain the following data. Set up and interpret a 90 percent confidence interval.

2.25	2.50
3.00	3.00
2.75	3.90
2.80	2.10
2.60	3.10

5. We wish to estimate how far the average suburban commuter drives to work in a large metropolitan area. We take a simple random sample of 17 drivers and obtain the following data (rounded to the nearest mile). Set up and interpret a 99 percent confidence interval for the unknown population mean commuting distance.

13	22	6	15	15
10	31	16	17	19
15	6	17	15	12
13	13			

6. Critique the following statement.

   The purpose of constructing confidence intervals is to estimate the sample mean with some degree of confidence.

7. A utility company has 75,000 customers. It hires a market research firm to randomly sample 1,000 households and to estimate the average amount of insulation in the 75,000 customers' homes. The average amount of insulation in the sample of households is ten inches. The sample standard deviation is 12 inches. Set up and interpret a 95 percent confidence interval.

8. The above utility company also wishes to know what proportion of the households have storm windows. Two hundred sixty households in the sample indicate that they have storm windows. Set up and interpret a 95 percent confidence interval of the

proportion of the 75,000 households in the population that have storm windows.

9. A curriculum review committee wishes to estimate the proportion of its recent graduates that found their statistics course fun and exciting. They take a simple random sample of 900 graduates and find that 50 found the course fun (this world is full of nerds and other weirdos). Set up a 95 percent confidence interval on the proportion of graduates who found statistics fun.
10. What proportion of ARC televisions have had five years of trouble-free operation? We take a simple random sample of 400 ARC owners and find that 300 have had no problems over the first five years. Set up and interpret a 90 percent confidence interval for the population proportion of ARC TV owners who have had no problems with their sets.

4:5 SOME FINAL THOUGHTS ON CONFIDENCE INTERVALS

By the end of this unit you should be able to:

1. Distinguish between the degree of certainty and meaningfulness in constructing confidence intervals.
2. Explain why it is impossible to construct a meaningful 100 percent confidence interval.
3. Explain the problems you would have in constructing confidence intervals if the central limit theorem did not exist.

Why settle for less than 100 percent certainty? Why not construct 100 percent confidence intervals? If you would rather be 95 percent confident than 30 percent confident, wouldn't you rather be 100 percent confident than 95 percent confident? The answer may surprise you.

Please construct an 80 percent, a 95 percent, and a 99 percent confidence interval for the following data. Then develop a rule that relates degree of confidence to margin of error. You'll need Table 4-2 on page 147 for your t values.

$$x\text{-bar} = 100 \quad n = 37 \quad s = 182.49$$

Here they are:

1. Eighty percent confidence interval: $100 \pm (1.306 \times 30)$ (60.82 to 139.18)
2. Ninety-five percent confidence interval: $100 \pm (2.030 \times 30)$ (39.10 to 160.90)
3. Ninety-nine percent confidence interval: $100 \pm (2.722 \times 30)$ (18.34 to 181.66)

The margin of error increases when you increase the degree of confidence. You would probably agree that wider confidence intervals provide you with less meaningful information. After all, if I am 95 percent confident that the average grade on the next exam will be between 10 and 90, have I given you meaningful information?

If you set up a 100 percent confidence interval your results will be worthless. Student's *t* table does not include the *t* value for a 100 percent confidence interval. Can you guess what the *t* value would be?

The *t* value for a 100 percent confidence interval is plus or minus infinity. The total area under Student's *t* curve is 100 percent. Fifty percent of the area lies between minus infinity and the mean of Student's *t* curve. In the above example we would be 100 percent confident that the unknown population mean would be between 100 minus infinity and 100 plus infinity. That information is worthless.

Basically you have two choices in constructing a confidence interval. Keeping the sample size constant, you can construct:

1. a relatively narrow interval with a low degree of confidence, or
2. a relatively wide interval with a high degree of confidence.

Ninety or 95 percent confidence levels are often used by companies. However, there is no magic number to use. Each person or company must decide upon the tradeoffs between degree of certainty and margin of error.

Actually, there are two ways to overcome the limitations of the two options offered above. How can you construct a relatively narrow 99 percent confidence interval? Before answering the question, look at the expressions for constructing intervals below.

$$x\text{-bar} \pm t \times s \div \sqrt{n}$$

$$p\text{-hat} \pm z \times \sqrt{p\text{-hat} \times (1 - p\text{-hat}) \div n}$$

One way is to increase the sample size. As the sample size increases, the standard error of the mean or proportion decreases due to the averaging-out effect. This leads to a smaller margin of error and a narrower confidence interval. Of course, this approach is costly because sampling is expensive and time-consuming.

The second way would be to reduce the standard deviation of the population. Unfortunately, in the short run, you cannot reduce it. It is the **inherent variability** in the population and it is difficult to reduce. However, we might be able to reduce it in the long run.

Imagine a machine that produces contact lenses. Although the lenses are supposed to be the same thickness, they are not: there is variability. It is due to the thousands of human and technological factors that cause the lens thicknesses to vary. In the short run, there is little we can do. However, in the long run we can purchase machines that produce more uniform lens thicknesses. We can purchase raw material from a supplier who adheres to better quality control standards. We can improve the training of the workers. All these factors will reduce the inherent variability within the population. Then the sample standard deviation which estimates sigma will also be smaller and we will get a narrower confidence interval.

Imagine that the central limit theorem, which states that the sampling distribution will be normal shaped provided the sample size is sufficiently large, did not exist. Would it have any impact on constructing confidence intervals? Why or why not? This is difficult, so take your time.

Yes, and here's the argument:

1. The central limit theorem tells us that the sampling distribution of the sample mean or proportion will be normal shaped.
2. Therefore, we can use the normal or Student's t (chubby normal) tables to find the z or t values.

Here's what would happen if the central limit theorem didn't exist:

1. There is no guarantee that the sampling distribution would be normal.
2. Then we could not use the normal or Student's t tables to construct confidence intervals.

4:6 HYPOTHESIS TESTING: AN ALTERNATIVE APPROACH TO CONFIDENCE INTERVALS

By the end of this unit you should be able to:

1. Differentiate between the confidence interval and hypothesis testing procedures.
2. Formulate hypotheses and their associated alternative actions.
3. Explain to others what Type 1 and Type 2 errors are.
4. Determine the costs associated with the two types of errors in a typical business decision.
5. Test the hypotheses and take action.

The goal in constructing confidence intervals is estimating unknown population parameters. Once constructed, we use the intervals to help us make decisions. Decision-making is better highlighted using the hypothesis testing framework. Here we run a study and assess the evidence provided by the data in favor of two claims or hypotheses concerning the unknown population parameter. Associated with each hypothesis is a different action.

We can test hypotheses about population means or proportions. In either case, the hypothesis testing framework involves five steps:

1. State the hypotheses and decision-making alternatives.
2. Determine the costs associated with the two types of decision-making errors.
3. Choose the significance level, alpha.
4. Collect the data and compute the sample statistics.
5. Test the null hypothesis and make your decision.

4:6:1 Testing Hypotheses About Unknown Population Means

I'll illustrate the hypothesis testing framework using Milemaster tires as an example. The company has decided that unless the average tire life of the Milemaster exceeds 47,000 miles, it is not worth marketing the product. The tread design engineering staff claims that the Milemaster will exceed the 47,000 mile figure. But you must make the final decision. We'll use the hypothesis testing framework to help you make a decision. To market or not to market, that is the question.

Stating the hypotheses. We begin by defining the **null hypothesis**. A hypothesis is a claim about the unknown population mean or proportion. The null hypothesis is a statement of "no effect" or "no difference." It is the hypothesis that says, "I know what you claim but I don't believe you." We also define an **alternative hypothesis** which says there is an effect or a difference. You also need to state what actions you will take if you accept the null or alternative hypothesis. These are the null and alternative actions. In the Milemaster decision you must decide between:

Null Hypothesis: the unknown population mean tire life is less than or equal to 47,000 miles

Null Action: If you accept (or fail to reject) the null hypothesis you will scrap the Milemaster tire project.

Alternative Hypothesis: the unknown population mean tire life is greater than 47,000 miles

Alternative Action: If you accept the alternative hypothesis you will market the tire.

Note several important features of the two hypotheses. First, the null hypothesis (H_0—read as H-nought) assumes that the average life of the Milemaster tire will not exceed 47,000 miles and should not be marketed. While the engineers claim that the tire will last more than 47,000 miles, you are the decision-maker and you must be *skeptical.* The null hypothesis reflects your skepticism. What you are saying is that, until otherwise proven, I will assume that the tire does not do what my engineering staff is claiming. Of course, you hope that based upon the sample evidence you will reject the null hypothesis and market the product. In general, the null hypothesis is the "I don't believe your claims" hypothesis while the alternative hypothesis (H_1) is the "I believe your claims" hypothesis. We assume the null hypothesis is true until otherwise proven.

Secondly, both hypotheses make claims about the unknown population mean. Unless you plan to sample the entire population, you need to draw inductive inferences. Hypothesis testing is an alternative approach to constructing confidence intervals.

Thirdly, the two hypotheses will always be mutually exclusive and exhaustive. Either the average tire life is less than or equal to 47,000 miles or it is more than 47,000 miles. Both hypotheses can't be true, but one must be true.

Determining the costs of the decision-making errors. You know that you can be either right or wrong when you draw inferences about unknown population parameters. The hypothesis testing framework illustrates that there are two ways to be right and two ways to be wrong when you draw inferences. I've shown this in the following table.

Based Upon the Sample Evidence You Decide to	The Truth About the Population Parameter	
	Null Is True	**Null Is False**
Accept the Null	correct decision	Type 2 error
Reject the Null	Type 1 error	correct decision

Type 1 error: You reject the null hypothesis when it is true.
Type 2 error: You accept the null hypothesis when it is false.

After you've collected the data and made your decision, only one of the errors is possible. You either accept the null hypothesis (in which case you've made a correct decision or a Type 2 error) or you reject the null hypothesis (in which case you've made either a correct decision or a Type 1 error). Unfortunately, before you collect the data you don't know whether you will accept or reject the null hypothesis. Therefore, you must be concerned about both errors and their associated costs.

The previous table is redrawn below so that it applies to the Milemaster tire decision.

Based Upon the Sample Evidence You Decide to	The Truth About the Population Mean Tire Life	
	47,000 Miles or Under	**Over 47,000 Miles**
Scrap the Tire	correct decision	Type 2 error
Market the Tire	Type 1 error	correct decision

When you make an error in business it costs you money. What are the costs associated with making a Type 1 error? You make a Type 1 error if you market the tire and the average life of the tire turns out to be 47,000 miles or under. Can you think of any costs associated with this Type 1 error?

When you advertise the tire you will claim that, on the average, it will last longer than 47,000 miles. In fact, the tire will not last more than 47,000 miles. You will not discover this overnight. You'll learn this only after a large number of Milemasters have been sold. That's when your problems will begin. Possible costs include reduced consumer confidence in your company, significant tire replacement costs under the tread warranty, and possible class action suits filed by irate consumers. In this example, a Type 1 error could be very costly. You should avoid it if possible.

What are the costs associated with a Type 2 error? Here you scrap the tire when in fact the average tire life would have been over 47,000 miles. You scrapped a superior tire; you may have lost a real marketing opportunity. The potential losses depend upon how good the tire was. If the average tire life was only 48,000 miles (only slightly better than currently available competitors' brands which have a 47,000 mile tread warranty), your losses might be small. But if the average tire life was 75,000 miles, then failure to market this tire could be very costly to your firm.

Setting a level of significance. No one likes to make errors. But in inductive inference there is always the chance you will make an incorrect decision. The level of significance is the *maximum* risk you are willing to accept in making a Type 1 error. It is called the **alpha level**. Clearly, the greater the costs associated with a Type 1 error, the lower you should set the maximum risk. It's your decision. Here are some guidelines you can use in setting the level of significance.

1. If the Type 1 error is potentially costly and the Type 2 error is not, set the level of significance very low—at .05 or less.
2. If the Type 1 error is not costly but the Type 2 error is, set the level of significance higher—perhaps at .25 and above.

3. If both errors are costly, set the level of significance very low and increase the sample size of your study. This reduces the chances of making a Type 2 error. However, larger sample sizes also are costly.

The idea behind the three rules is to minimize your chances of making errors that can cost you your job. Textbooks commonly stress certain standard levels of significance such as five percent or one percent. It makes no sense to treat five percent as a universal standard. It should depend upon the costs associated with making a Type 1 error. That's what the guidelines are for.

In the Milemaster study, the *potential* costs associated with a Type 1 error are very large. Therefore, I recommend setting the level of significance very low—at one percent. This means that if you reject the null hypothesis, the maximum chance of making a Type 1 error is one in a hundred. In other words, you can be at least 99 percent confident that you have made a correct decision.

Data collection. Design your study and collect your data. Let's suppose you decide to take a simple random sample of 1,000 tires from the population. You run the tires until they fail to meet federal guidelines for tread depth. The sample mean and sample standard deviation are

$$x\text{-bar} = 48{,}050 \text{ miles} \qquad s = 3{,}162.27 \text{ miles.}$$

Hypothesis testing and inductive inference are only valid for properly designed studies. No amount of analysis can correct basic flaws in the design or conduct of a study. I've already presented one strategy for designing valid studies in conjunction with confidence intervals, namely, *use probability sampling*. Select your tires, your test personnel, and your test equipment at random. In the next chapter I will present several other rules of good experimentation.

Weighing the evidence and making your decision. Do the sample statistics, *x*-bar and *s*, favor the null or the alternative hypothesis? In our example, the two hypotheses were:

H_0: the population average tire life is less than or equal to 47,000 miles

H_1: the population average tire life is greater than 47,000 miles

You selected only *one* sample of 1,000 tires. However, if you had selected *all possible* samples of 1,000, you could have constructed a sampling distribution for the mean. It is drawn on page 165 *under the assumption that the null hypothesis is true*. Remember that we assume the null hypothesis is true until proven otherwise. The mean of the sampling distribution would be 47,000 miles (which is the most the population mean could be under the null hypothesis) and the estimated standard error is 100 miles (or $3{,}162.27 \div \sqrt{1{,}000}$). The shape of the sampling distribution is normal due to the central limit theorem.

Figure 4-8 Sampling Distribution—Assuming the Null Hypothesis Is True

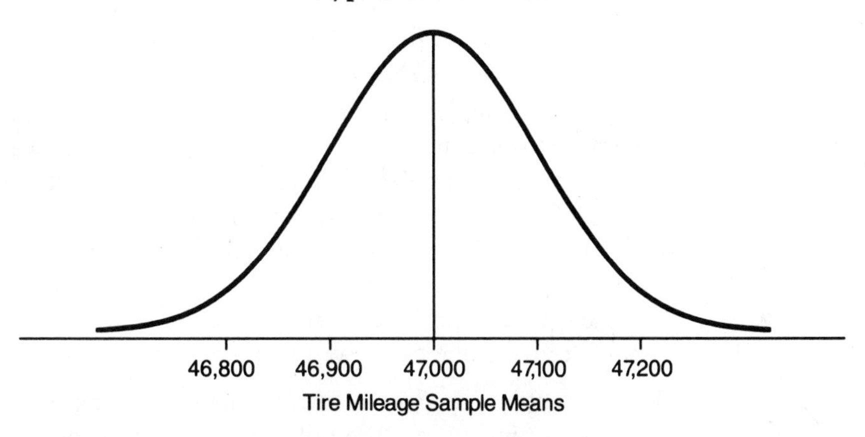

Does the sample mean of 48,050 miles favor the null or alternative hypothesis? Based upon the sample mean should we accept or reject the null hypothesis? Before answering that question, please answer the following one. If the sample mean had equalled 22,000 miles would you accept or reject the null hypothesis and why? That is, is a sample mean of 22,000 miles more likely if the unknown population mean tire life was at most 47,000 miles or greater than 47,000 miles?

I hope you said that you would accept (or fail to reject) the null hypothesis. A sample mean of 22,000 miles is more likely to occur when the unknown population mean is less than or equal to 47,000 miles than when it is greater than 47,000 miles.

If the sample mean had equalled 90,000 miles wouldn't you have concluded that the population mean is more likely to be over 47,000 miles? Thus as the sample mean increases the evidence begins to favor the alternative hypothesis. How far above 47,000 miles must the sample mean be before you should reject the null hypothesis? That depends upon the level of significance that you choose.

In this example we set the level of significance, alpha, at one percent. This is equal to the area that I've shaded in the sampling distribution in Figure 4-9. This area is called the **region of rejection**. If the sample evidence (*x*-bar) falls in this region you will reject the null and market the Milemaster. If the sample mean falls to the left of the shaded area you will accept the null and scrap the Milemaster project.

Here's the logic. If the sample mean falls in the region of rejection it could only be due to one of two possibilities.

1. The null is true. Your chances of getting a sample mean in the region of rejection if the null is true is at most one in a hundred.
2. The null is false. The reason that your sample mean was so large (so that it fell in the shaded area) is that the null hypothesis is incorrect. The alternative hypothesis which states that the population mean is greater than 47,000 miles is true.

Which explanation is more reasonable? The chances of getting a sample mean in the region of rejection if the null is true is at most one in a hundred. Therefore the second explanation is more reasonable. You should conclude that the null hypothesis is false.

Figure 4-9 Sampling Distribution and the Region of Rejection

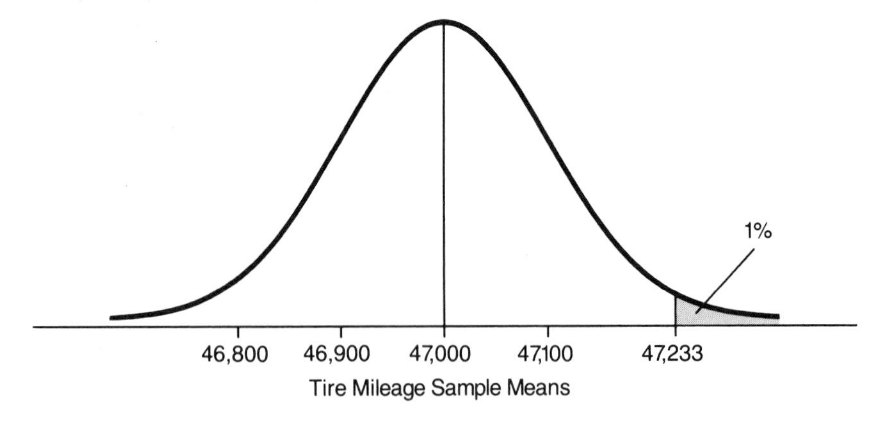

The value that separates the acceptance and rejection areas under the curve is called the **critical value**. Here's how you compute it:

$$\text{critical value} = 47{,}000 + t \times 100 \text{ miles}$$

You use the Student's t value because you don't know the population standard deviation. However, in this example the t and z values are identical because the sample size is 1,000 tires. If the shaded area under the curve is one percent, then the area between the mean and the critical value is 49 percent. The z value associated with 49 percent of the area under the normal curve is 2.33 (see Table 4-1 on page 145). The critical value is:

$$47{,}000 + z \times 100$$

or

$$47{,}000 + 2.33 \times 100 = 47{,}233 \text{ miles}$$

This leads to these two decision rules.

If the sample mean for the 1,000 tires is less than or equal to 47,233 miles, you will accept the null hypothesis and scrap the project.

If the sample mean for the 1,000 tires is greater than 47,233 miles, you will reject the null hypothesis and market the Milemaster tire.

Since the sample mean, based upon 1,000 randomly selected tires, was 48,050 miles, you reject the null hypothesis. You would market the Milemaster tire.

At this point you have either made a correct decision or you have rejected the null when you shouldn't have—a Type 1 error (see the table on page 162). You cannot have made a Type 2 error. Making a Type 2 error means you accept the null hypothesis when it is false. But you didn't accept the null, you rejected it.

The actual chance of making a Type 1 error is called the **prob-value** or ***p*-value**. The p-value is the likelihood of getting a sample mean of 48,050 miles or larger if the null hypothesis were true. It's shown in Figure 4-10 on the next page.

Visually, the p-value is the area underneath the curve to the right of the actual sample mean of 48,050 miles. In the example the p-value is much less than one percent. The p-value tells you the *actual* probability that you made a Type 1 error, given the sample results you obtained. The level of significance is the *maximum* probability that you are willing to tolerate of making a Type 1 error.

Figure 4-10 Right-Hand Tail of the Sampling Distribution and the *p*-Value

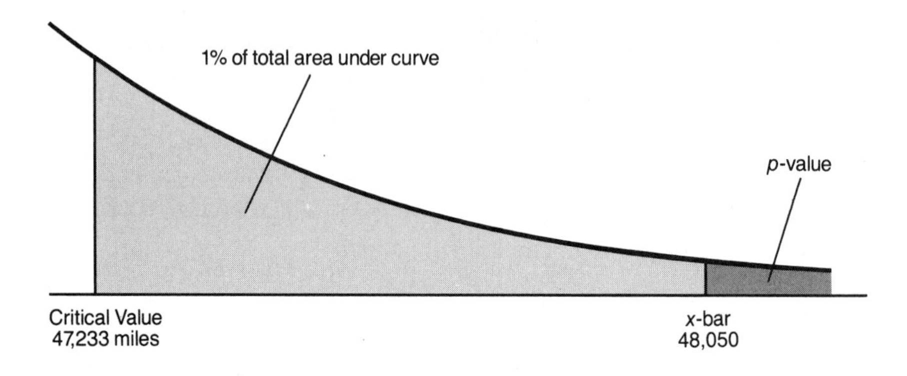

In conclusion, based upon the sample evidence, we have rejected the null hypothesis and will market the Milemaster. Since we based our decision on a small sample, we can be wrong. However, the chance that we made a Type 1 error is much less than one in a hundred. We can be very sure that we made a correct decision.

4:6:2 Testing Hypotheses About Unknown Population Proportions

You are the owner of a chain of personal computer stores and are considering the purchase of a Quiklearn training system. You've found that you can increase your sales of microcomputers by offering quality hands-on training in programming. Quiklearn is offering a system that consists of ten self-paced lessons in BASIC programming. They claim that over 70 percent of people with no prior computer experience who go through the lessons learn successfully to program in BASIC. If this claim is true, you will purchase their system. Before making the final decision about the Quiklearn training method, you want to test their claim.

Stating the hypotheses. The general rule is that your null hypothesis is the "I don't believe what you are telling me" hypothesis. With this in mind, what are the the null and alternative hypotheses and actions in plain English? Remember that Quiklearn is making a claim about the unknown population proportion, not the population mean. Please don't look ahead.

Hypothesis	Action
H_0: Seventy percent or fewer of the customers who take the lessons will successfully learn BASIC programming.	Don't purchase the Quiklearn system.
H_1: Over 70 percent will successfully learn BASIC programming.	Purchase the Quiklearn system.

Quiklearn may claim that their system can teach BASIC in ten lessons but you must not assume that their claim is true. Remember, be skeptical of others' claims.

Determining the costs of the decision-making errors. Here is what goes wrong if you make each error.

Error	What Goes Wrong
Type 1: You reject the null hypothesis when it is true.	You purchase the training method when in fact 70 percent or fewer of your customers will learn BASIC.
Type 2: You accept the null hypothesis when it is false.	You don't purchase the training when over 70 percent would have learned BASIC.

In the space provided, list the costs associated with making each error. Be specific.

Type 1 Error Costs	Type 2 Error Costs
1. cost of the purchase 2. cost of implementing the training in your chain 3. angry customers who thought they would learn BASIC in ten lessons and have not 4. lost computer sales due to angry customers	1. lost sales because you don't have a good BASIC training method to offer your customers, while your competition does.

Setting the level of significance. Given all the potential costs, you set the level of significance at 15 percent. This is the maximum risk you are willing to tolerate of making a Type 1 error. In other words, if you implement the training system, you want to be at least 85 percent sure that over 70 percent of your customers who take the lessons will be able to program in BASIC.

The way you plan to test the null hypothesis is to try the system on a sample of 100 randomly selected customers and see what percentage successfully learn BASIC. With a sample size of 100, the sampling distribution (a binomial distribution) will be normal shaped. Assuming that the null hypothesis is true, the center of the normal curve is the assumed population proportion of .70, and the standard error is

$$\sqrt{\frac{p \times (1 - p)}{n}} = \sqrt{\frac{.70 \times .30}{100}} = .046$$

With this information we can draw the sampling distribution for the population proportion assuming the null hypothesis is true.

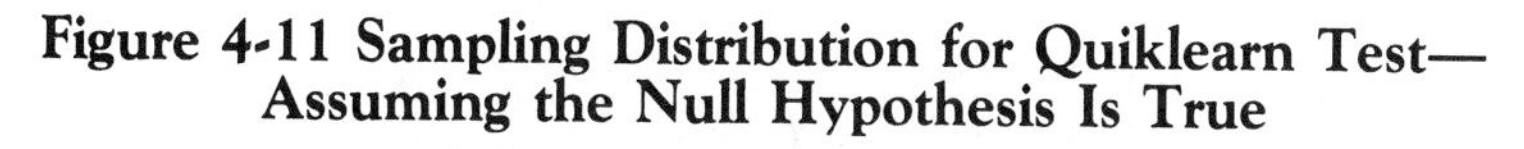

Figure 4-11 Sampling Distribution for Quiklearn Test—Assuming the Null Hypothesis Is True

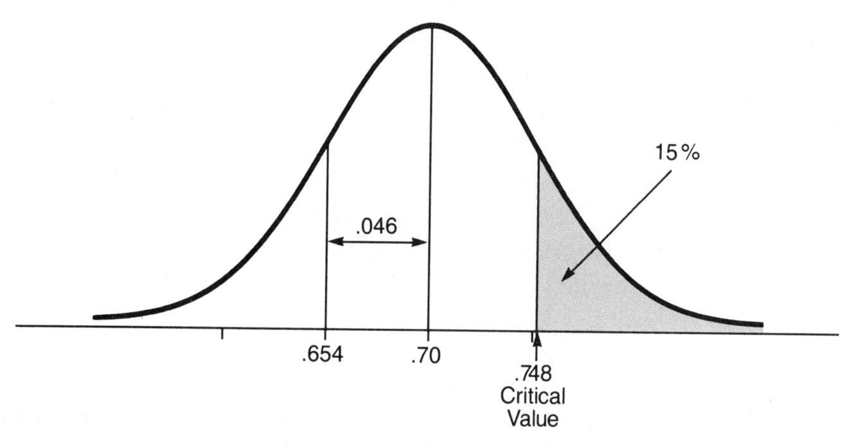

The region of rejection will be in the right-hand tail of the distribution. Why? A sample proportion in the left-hand tail (say, .30 or .40) would clearly be consistent with the null hypothesis. Sample proportions much larger than .70 would favor the alternative hypothesis. Therefore, the critical value must lie in the right-hand tail.

The critical value which separates the regions of acceptance and rejection depends upon the level of significance you set. In this example, 15 percent of the area under the sampling distribution lies to the right of the critical value. Because the sampling distribution is normal shaped, we use the z value from Table 4-1 on page 145. The z value is 1.038. The critical value is equal to:

$$.70 + z(35 \text{ percent}) \times \text{standard error}$$

or

$$.70 + (1.038 \times .046) = .748$$

Here are the decision rules.

If the proportion of the 100 customers who successfully learn BASIC is less than or equal to .748 you will accept the null hypothesis and not purchase the training method.

If the sample proportion from the study is greater than .748, you will reject the null hypothesis and purchase the training method. In other words, you will reject the null hypothesis if 75 or more customers successfully learn BASIC in the test program.

Data collection. You test the Quiklearn method on 100 randomly selected customers before making your final decision. After each customer completes the lessons you determine whether they have learned BASIC programming. The result: 71 out of 100 in the sample successfully learned BASIC from the Quiklearn lessons.

Weighing the evidence and making your decision. Since the sample proportion ($^{71}/_{100}$) falls in the acceptance region you will not purchase the Quiklearn training method. You have just made either a correct decision or made a Type 2 error (see page 162). Since you failed to reject the null hypothesis, you cannot have made a Type 1 error. It's possible that you made a Type 2 error. However, determining the probability of making a Type 2 error is far beyond the scope of this book.

There is much more to hypothesis testing. However, my objectives for this section were limited. The point was to demonstrate the differences between hypothesis testing and confidence intervals. Hypothesis testing is very decision-oriented and does illustrate that two different types of errors can be made in inductive inference.

Before leaving this chapter, here are the most important ideas.

1. You draw inductive inferences when you take probability samples from a population, compute sample statistics, and use them to either construct confidence intervals or do hypothesis testing.
2. You reduce the chances of selection bias by taking probability samples from the population. A probability sample requires that each object in the population has a chance of being in the sample.
3. The standard error of the sample mean (the standard deviation of the sampling distribution) is always less than the population standard deviation. Cost considerations aside, you should always take the largest sample size possible to estimate a population parameter.
4. Regardless of the shape of the population, the central limit theorem assures us that the sampling distribution will be normal shaped. This allows us to use the normal or Student's *t* curves in computing the margin of error in confidence intervals or the critical value in hypothesis testing.
5. When we don't know the population standard deviation, we use Student's *t* curve and not the normal curve to compute the margin of error. This will widen out the confidence interval.

6. The hypothesis testing framework highlights decision-making better than the confidence interval approach.
7. There are two types of errors you can make in inductive inference. You must determine their respective costs and set the level of significance for your study consistent with the cost of making the Type 1 error. The higher the potential cost, the lower you should set the level of significance.

Exercise Set for 4:6

1. An engineering group at an automotive manufacturer claims that its new brake system will stop a car going 30 miles per hour in under 60 feet. This is much better than the present brake system. You must decide whether to approve its use in the upcoming year's model. Set up the null and alternative hypotheses and actions. Determine the economic costs associated with making a Type 1 and Type 2 error. The data from the study are x-bar = 58 feet, s = ten feet, and $n = 100$. Use a level of significance of ten percent. What decision should you make? Show all work.
2. A company is considering purchasing robots, for a welding operation. In order to cost justify the robots, they must produce, on the average, more than 20 good welds per minute. You must make the final purchasing decision. Set up the null and alternative hypotheses and actions. What are the economic costs of making a Type 1 and Type 2 error? The data from the study are x-bar = 20.2 good welds per minute, $s = 7.07$ good welds per minute, and $n = 100$. Given a level of significance of one percent, what decision should you make?
3. In plain English, what is the difference between level of significance and p-value?
4. Is it possible that once you make your decision, you can make a Type 1 and Type 2 error? Discuss.
5. In the American legal system you are assumed to be innocent until proven guilty. The jury must decide the person's innocence or guilt on the basis of incomplete evidence (sample evidence). Set up the null and alternative hypotheses and actions. What are the social costs of making a Type 1 and Type 2 error?

6. Why was it necessary to use inductive inference in Problem 2? Since the sample mean was 20.2 good welds per minute, which is greater than 20, why couldn't we reject the null hypothesis? Why did we have to use inductive inference (determine the critical value and compare the sample mean to it)?
7. From past records, a computer leasing firm knows that at least 10 percent of all customers are not satisfied with the service on their leased computers. Recently the firm launched a new customer service program to reduce the percentage of dissatisfied customers. Now they want to see if it's working. Set up the null and alternative hypotheses for the study. The firm samples 200 customers and only 8 of them are now dissatisfied. Given this data should the firm permanently adopt the new customer service program? Is it reducing the population proportion of customer complaints? Use a 10 percent level of significance.

5. Experimentation and Analysis of Variance

5:1 INTRODUCTION

In the last chapter we obtained data by sampling procedures. We took simple random samples from a population, tested or observed certain characteristics, and computed sample statistics. We used the statistics to make inductive inferences.

Experimentation differs from sampling in that we *vary* certain factors and test or observe their effects. We compute statistics and make inductive inferences based on these effects. For instance, what happens to problem-solving ability as the amount of sleep is decreased? We randomly select subjects, vary the amount of sleep that each is permitted, and then determine the impact on problem-solving ability.

We use sampling to estimate the problem-solving ability of the population. We use experimentation to determine what factors improve problem-solving ability. Sampling answers the question "how much?" and experimentation answers the question "what causes?"

Here is some important terminology.

Experimental units are the units on which the experiment is done. When the units are people, we call them **subjects**.

The **experimental treatment** or **factor** is what is varied; for example, the amount of sleep. We draw inductive inferences on the impact of the factor on the dependent variable.

The **dependent variable** is what is measured as we vary the factor (for example, problem-solving ability).

Experimentation is similar to sampling in two respects. First, we use inductive inference procedures for both. We take small samples and draw inductive inferences to the population. Second, in sampling, if we select our subjects or phrase the questions improperly, the confidence interval has no meaning. Similarly, if we don't conduct an experiment properly, the analysis has no meaning. Later I will present some general principles of good experimentation. But first you must learn how to analyze the experimental data for a one factor completely random design.

5:2 THE ONE FACTOR COMPLETELY RANDOM DESIGN

By the end of this unit you should be able to:

1. Visually inspect a data set and determine with 80 percent accuracy whether or not the factor affects the dependent variable.
2. Explain how the sum of squares decomposition works without resorting to statistical jargon or formulas.
3. Design a valid one factor completely random experiment.

We have just completed running an experiment in which three brands of gasoline (the single factor) have been tested to determine which brand, if any, delivers the highest miles per gallon (the dependent variable). The experiment was conducted properly. One driver and one car (experimental units) were used in the study and all the test runs were made under the same track conditions. We continually checked the car to ensure that it was properly maintained. We also rested the driver to ensure that he did not become fatigued during the test runs.

We tested each brand of gasoline five times (five tankfuls). Thus it will be necessary to draw inductive inferences as to which brand of gasoline, if any, will deliver the highest miles per gallon (mpg) in the *long run*. After all, if a brand is best in the study we want to be able to say that it will be best for any car and any driver.

Just as in hypothesis testing, we must state a null and alternative hypothesis.

Null hypothesis: There is *no* difference in the long-run mpg of the three brands of gas.

Alternative hypothesis: There *is* a difference.

Here are two different data sets that might have resulted from our test.

	Data Set 1			Data Set 2		
	Brand 1	Brand 2	Brand 3	Brand 1	Brand 2	Brand 3
	20	25	28	18	27	17
	22	27	28	24	20	37
	21	26	27	17	29	29
	22	26	29	22	31	21
	20	26	28	24	23	36
Average	21	26	28	21	26	28

Both data sets have the same miles per gallon sample averages (Brand 1 = 21, Brand 2 = 26, Brand 3 = 28). But note that the spread in the miles per gallon figures within each of the brands in Data Set 1 is much less than for the brands in Data Set 2. When we draw our inductive inferences on the two data sets, we will find that there really is a long-run difference in the mileage figures for only one of the data sets. Without doing any mathematics, can you determine which is which and why? If you can't yet, don't be discouraged because by the end of this unit you will be able to do it.

If you determined that for Data Set 1 we would find a long-run difference (reject the null hypothesis) in mpg due to brands of gasoline, you are right. Before discussing the logic behind that conclusion, I'll show you a way to come to the same conclusion using pictures.

5:2:1 A Quick Visual Analysis for a One Factor Completely Random Design

Let's draw some pictures that represent the two data sets. I call these **spread charts**. The numbers in the spread charts represent the brands of gasolines. They're arranged along the number line according to the mileage obtained in each test run.

Figure 5-1 Spread Charts

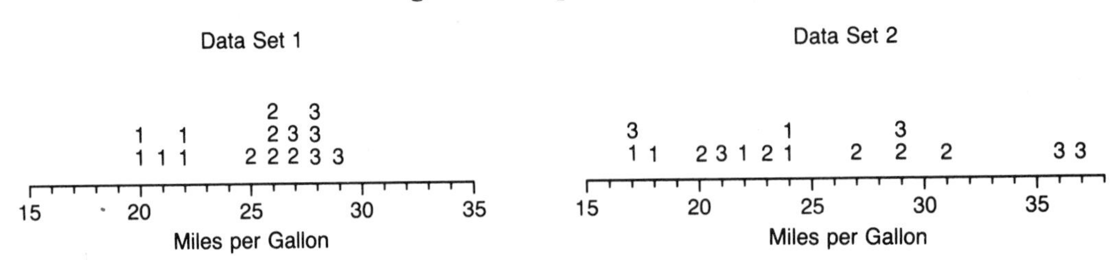

Notice that in Data Set 1 the mileage figures for each brand of gasoline cluster together; there is little or no overlap between the figures for the three brands. This is not true for Data Set 2. The observations for the three brands of gasoline all overlap with one another. What does that tell us?

The average miles per gallon figures in Data Set 1 appear to be *stable*; there is very little spread within each brand. Even if we made several more runs on each brand of gasoline, we would not expect the three sample averages to change very much from the present values. The sample means in Data Set 1 will tend to maintain their rank order (Brand 1 the least and Brand 3 the most mileage). Thus we conclude that there is a long-run difference between the three brands. We reject the null hypothesis. The brands appear to affect gas mileage.

The average miles per gallon figures are *unstable* in Data Set 2. Suppose we make two more runs with each brand of gasoline in Data Set 2 and recompute the three sample averages. Would the new miles per gallon figures be very close to the present averages of 21, 26, and 28 mpg? Since the sample averages appear to be unstable, they could change dramatically. Brand 1 which now has the lowest sample average could be the highest average after seven runs. Thus we cannot be sure if the differences between the three sample means after only five observations indicate a real long-run, or population, difference in miles per gallon. We cannot reject the null hypothesis. We can't be confident that the brands affect gas mileage.

You should get into a habit of drawing spread charts of the data before you begin your formal analysis. It will help you get familiar with the data before massaging the numbers. Remember this is not a substitute for a formal statistical analysis but it will help you comprehend what the formal analysis is actually doing. Let's see how well you can do.

Below are two additional data sets. Do the three machines have the same long-run productivity levels? Do the three merchandising displays produce the same long-run revenue? For which data set are you likely to reject the no long-run difference null hypothesis? Please draw spread charts for the two data sets and make an educated guess.

	Data Set 3			Data Set 4		
	Machine 1	Machine 2	Machine 3	Display 1	Display 2	Display 3
	50 units	60 units	70 units	$180	$218	$210
	52 units	58 units	69 units	$200	$178	$215
	50 units	62 units	71 units	$200	$200	$170
	48 units	60 units	70 units	$220	$196	$213
Average	50 units	60 units	70 units	$200	$198	$202

I hope you concluded that for Data Set 3 you would *probably* reject the null hypothesis and conclude that not all three machines will have the same productivity in the long-run. The word "probably" is necessary because without formal statistical analysis we cannot even assign probabilities to our inductive inferences. Remember we are drawing inferences from this one experiment to the long run or population.

Here's the logic behind that conclusion. Consider the amount of overlap between the three columns of data in Data Set 3. Note that they do not overlap with one another (48–52 units for Machine 1; 58–62 units for Machine 2; and 68–72 units for Machine 3). This suggests that if we were to continue running the experiment, observations for Machine 1 would almost always be smaller than for Machine 2, and observations for Machine 2 would almost always be smaller than for Machine 3. That suggests that there is a long-run difference in productivity.

You can also analyze the spread *within* and *between* the levels of the factor. The spread in the productivity data values for each machine is very small; in the spread chart there are three non-overlapping data clusters, one for each level of the factor. Also there appears to be a large difference in the three sample means. This suggests that there is a long-run difference in the productivities of the three machines.

In Data Set 4 it is impossible to draw three non-overlapping data clusters for the three displays. There is a large amount of spread in the data values *within* each level of the factor. Further, there is little spread *between* the three sample averages. The closer the sample averages are to one another, the more difficult it is to reject the null hypothesis. If we continued to run the study, the three sample means might change from their present rank ordering. We can't conclude that there is any long-run difference in sales revenue between the three displays.

Please develop a rule of thumb for determining when an experimental factor does or does not affect the dependent variable. Base it on the amount of spread in the observations within and between the levels of an experimental factor. What are your conclusions for each of the four situations below? Please write them down before looking ahead.

Spread *Between* the Three Sample Averages	**Spread *Within* the Three Sets of Observations**
1. little	little
2. little	much
3. much	little
4. much	much

Do your conclusions agree with mine?

Spread *Between* the Three Sample Averages	Spread *Within* the Three Sets of Observations	Conclusions
1. little	little	Hard to tell
2. little	much	Factor has no impact; don't reject null hypothesis
3. much	little	Factor has an impact; reject the null hypothesis
4. much	much	Hard to tell

Data Set 3 is an example of the third condition. Data Set 4 is an example of the second condition. The other two conditions are more difficult to assess without a formal analysis. Now unless you could explain the rules of thumb in plain English to a friend, you still haven't mastered the idea. Of course, this quick technique is not a substitute for a formal statistical analysis. That's our next topic.

5:2:2 An Intuitive Sum of Squares Decomposition for the One Factor Completely Random Design

Let's return to Data Set 1 and formally analyze the data. The formal analysis begins with the *decomposition of the sum of squares*.

Data Set 1

	Brand 1	Brand 2	Brand 3
	20	25	28
	22	27	28
	21	26	27
	22	26	29
	20	26	28
Average	21 mpg	26 mpg	28 mpg

Overall Average = 25 mpg

The only formula we will need for the entire chapter is the basic definition of variance. Variance is "almost the average" of the squared differences of a set of observations around its mean. The formula for variance is

$$[(x_1 - x\text{-bar})^2 + (x_2 - x\text{-bar})^2 + \ldots + (x_n - x\text{-bar})^2] \div (n - 1)$$

The numerator is called the **sum of squares**. It is the sum of the squared differences of each observation about its mean. It is always equal to the sum of squared differences between the relevant observations and their mean. Think of sum of squares as units of variation in a data set. The denominator is called the **degrees of freedom**. It is always equal to the relevant sample size minus one. My use of the term "relevant" is explained below.

What is the total sum of squares in the 15 observations within Data Set 1? Since we are interested in 15 observations, the **relevant** mean is the mean of all 15 numbers (or the overall mean). Thus the sum of squares-total (SST) is the sum of the squared differences between each of the 15 observations and the overall mean.

$$\text{SST} = (20 - 25)^2 + (22 - 25)^2 + \ldots + (29 - 25)^2 + (28 - 25)^2$$

$$= 138 \text{ units of variation}$$

Without looking ahead, what two sources could account for the 138 units of variation? In other words, what are two reasons that the fifteen numbers are not all the same? This is what the sum of squares decomposition answers.

One source is the impact of the three brands of gasoline (or in general, the experimental factor). This is measured by the **sum of squares due to the treatment (SSTR)**. The other source is the impact of everything else. These are the literally thousands of other factors we could have experimented on but didn't. These are called **extraneous factors**.

We controlled some extraneous factors such as driver fatigue and general maintenance of the test car by constantly resting the driver and monitoring and adjusting the car. However, even in a well-controlled experiment, it is impossible to control *all* extraneous factors. How about time of day when we ran each of the brands? It is possible that during the day the wind conditions changed which could affect gas mileage. It is possible that the test track conditions changed throughout the day which could also affect gas mileage. The impact of the extraneous factors on miles per gallon is called the **sum of squares due to extraneous factors (SSEF)**.

The **sum of squares decomposition** tells you how much of the total variation in all the observations is due to the experimental factor and how much is due to all other extraneous factors.

$$SST = SSTR + SSEF$$

First let's compute the SSEF. Remember that the sum of squares is the sum of the squared differences between each relevant observation and its mean. Ask yourself the following question. Why is it that when Brand 1 gasoline was used, the five mpg figures were not all the same? Please insert your answer below.

Does it make sense that the differences must be due to the impact of the thousands of extraneous factors we chose to ignore in the study? So the relevant data are the five observations under Brand 1 and the relevant sample average is the average of those five numbers. This is also true for the data for Brand 2 and Brand 3. Thus we compute sum of squares due to extraneous factors as follows.

$$\begin{aligned} \text{SSEF} &= (20-21)^2 + \ldots + (20-21)^2 && \leftarrow \text{for Brand 1} \\ &\quad + (25-26)^2 + \ldots + (26-26)^2 && \leftarrow \text{for Brand 2} \\ &\quad + (28-28)^2 + \ldots + (28-28)^2 && \leftarrow \text{for Brand 3} \\ &= 8 \text{ units of variation} \end{aligned}$$

Sum of squares–extraneous factors accounts for eight out of the 138 units of total sum of squares. Thus the sum of squares due to the treatment (SSTR) must be 130 units. There is a direct way to calculate this number. We'll tackle that next.

What three numbers best represent the long-run miles per gallon of the three brands of gasolines? Hopefully, you said the three sample averages. These become the relevant observations in our sum of squares calculation. The mean of these three observations is merely the overall mean of 25 mpg (this is only true when the number of observations under each treatment level is the same). Given the basic definition, you are probably thinking that the sum of squares–treatment is:

$$(21-25)^2 + (26-25)^2 + (28-25)^2 \text{ or } 26 \text{ units}$$

That's a reasonable guess but it's not exactly correct. We know that the actual answer is 130 units of variation. We must multiply the above expression by five to get the correct value.

$$\begin{aligned} \text{SSTR} &= 5[(21-25)^2 + (26-25)^2 + (28-25)^2] \\ &= 130 \text{ units} \end{aligned}$$

The 5 is the number of observations under each brand of gasoline. This is the **weighting constant** and must be included when calculating the sum of squares–treatment. If we had run 5,000 observations for each brand of gasoline, the weighting constant would have been 5,000.

We have completed the sum of squares decomposition but not the analysis. Remember we are conducting an analysis of *variance*. All we need to do is divide the sum of squares by their relevant degrees of freedom to compute the variances. We then compare them and the analysis will be complete.

Why do the degrees of freedom equal $n - 1$? Suppose you are asked to compute the sum of squares–extraneous factors for three numbers whose average equals two. Could you correctly guess what the three numbers were if you were given no hints? Unless you are a psychic you couldn't. What if you were told that the average equalled two and one of the numbers was one? You still couldn't do it. What if you were told that the average was two and two of the numbers were one and five respectively? Could you determine the missing third number?

When the average and two of the three numbers are known, the third number can be determined and the sum of squares can be calculated. The third number must be zero if the average equals two. Thus when you know two (or $n - 1$) of the three numbers and the average, you can determine the final observation. In our example you could assign any values to the first two numbers (your choice—your freedom), but once that is done the third number is automatically determined (you've lost your freedom).

Remember, the number of degrees of freedom is always the relevant sample size minus one ($n - 1$). What are the degrees of freedom for the sum of squares–total? Since there is a total of 15 observations or calculations, the relevant sample size is 15. Therefore, there are 14 degrees of freedom.

What are the degrees of freedom for sum of squares–treatment? This term captures the impact of the three brands of gasoline. Therefore the relevant sample size is three and there are two degrees of freedom.

Now you can determine the degrees of freedom for the extraneous factors term. There are five observations under Brand 1. This generates four degrees of freedom. There are also four degrees of freedom each for Brand 2 and Brand 3. This gives 12 degrees of freedom. The results of the decomposition can be displayed in an **analysis of variance** (ANOVA) table.

ANOVA Table

Sources of Variation	Sum of Squares	Degrees of Freedom	Variance
Treatment	130	2	65
Extraneous	8	12	.666
Total	138	14	

You can see that the variance due to the treatment (the three brands of gasoline) is much greater than the variance due to all extraneous factors not controlled in the experiment. How much larger is it? We'll compute a ratio of the two variances. The variance ratio equals 65 ÷ .666, or 97.59. This tells us that the variance in the 15 mileage figures generated by the three brands of gasoline is almost 98 times larger than the variance generated by the extraneous factors alone. That seems to suggest that the brands of gasoline affect the mileage figures; we should reject the null hypothesis.

How large must the variance ratio be before you should reject the null hypothesis? To answer this question we will need to resort to inductive inference and the Fisher distribution.

Remember that if the number of observations is different for each level of the experimental factor, you will have to use more complicated formulas. Nevertheless, the logic of the decomposition will not be affected.

5:2:3 The Fisher Distribution

Why do we need to draw inductive inferences using the Fisher distribution? Why can't we just look at the variance ratio and determine if we should reject the null hypothesis? This is very important so jot down some reasons before reading on. (Recall the lessons of Chapter 4.)

In the business world we are interested in drawing conclusions that are valid for the long-run and not just for a particular experiment. Which gasoline brand, if any, will be the best performer? Not just in the experiment, but when we use it with our fleet of cars. Five observations under each brand of gasoline do not constitute the long-run. Thus our procedure is to run valid and controlled experiments and, based upon the sample statistics, draw an inductive inference about the long-run (or to the population).

From Chapter 4 you know you can be wrong when you draw inductive inferences. However, what's the alternative? You could run an infinitely long experiment which would eliminate the need for drawing inductive inferences. However, you would then substitute inaction for uncertainty. I would rather take an action that could be wrong rather than no action at all while awaiting the results of an experiment without end.

Statistics textbooks will tell you to compare the value of the variance ratio to the tabled values of the Fisher distribution (F values). If the variance ratio is larger than the F value, reject the null hypothesis, otherwise don't. Here's the reasoning behind that procedure.

The Fisher distribution is the sampling distribution for the variance ratio (and you thought you just needed to know sampling distributions in Chapter 4). We pretend to run an experiment such as the gasoline brand study not once, but an infinite number of times. Further we assume that there really isn't any long-run difference in the mpg of the three brands; we assume the null hypothesis is true. Imagine that the first experiment has been concluded and the variance ratio has been calculated. If there really isn't any long-run difference in the mpg for the three brands of gasoline, the value of the variance ratio should be approximately one. Why?

If the different brands of gasoline do not affect the long-run mpg, then this factor is no more important than all the extraneous factors we chose to ignore in our study. Thus the variance-treatment should be about the same as the variance–extraneous factors. The variance ratio would thus be close to one.

Here's another way to think about the two variance terms. The variance–extraneous factors measures the impact of all the extraneous factors the researcher chose to ignore. The variance-treatment measures the impact of all extraneous factors *plus* the impact of the brands of gasoline—the experimental factor. If the null hypothesis is true and the brands have no impact, then the numerator and denominator of the variance ratio will be equal to each other. Thus the variance ratio will equal one.

Each time we rerun this hypothetical experiment, we would expect the variance ratio to be close to one. Occasionally the variance ratio could be either smaller or larger than one. If we were to plot the distribution of variance ratios we would obtain a Fisher distribution. The Fisher distribution can take on values from zero (variances can never be negative!) to infinity and is shown below.

Figure 5-2 Fisher Distribution for 2 and 12 Degrees of Freedom

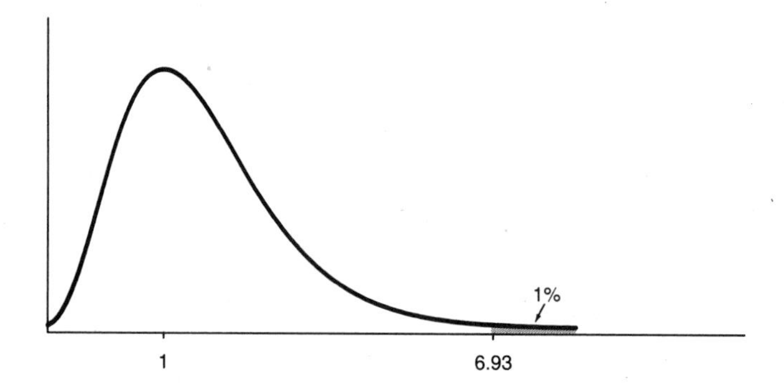

Like the Student's *t* curve, the Fisher curve is really a family of curves. The total area underneath the curve is 100 percent. Below is an abbreviated Fisher table.

A Short Fisher Table

df for Denominator	Degree of Confidence	df for Numerator 1	2	3
4	90 percent	4.54	4.32	4.19
	95 percent	7.71	6.94	6.59
	99 percent	21.20	18.00	16.69
6	90 percent	3.78	3.46	3.29
	95 percent	5.99	5.14	4.76
	99 percent	13.74	10.92	9.78
9	90 percent	3.36	3.01	2.81
	95 percent	5.12	4.26	3.86
	99 percent	10.56	8.02	6.99
12	90 percent	3.18	2.81	2.61
	95 percent	4.75	3.89	3.49
	99 percent	9.33	6.93	5.95

The Fisher value for 2 (degrees of freedom for the numerator of the variance ratio), 12 (degrees of freedom for the denominator), and a 99 percent level of confidence is 6.93. The area to the right of a variance ratio of 6.93 is one percent. This means that if there really was no long-run difference in the mpg of the three brands, only one percent of the time would we obtain a variance ratio larger than 6.93. In our actual study the variance ratio equalled 97.59. There are only two explanations for the very large variance ratio.

1. The null hypothesis is *true*. While the chance of obtaining a variance ratio of 97.59 is extremely rare, this is one of those times. If the chance of obtaining a variance ratio of 6.93 or larger is only one percent, the chance of obtaining a value of 97.59 or larger must be extremely small (less than one in ten thousand).
2. The null hypothesis is *false*. That is the reason we obtained such a large variance ratio.

I hope you think that the second explanation is more reasonable. You should reject the null hypothesis and conclude that the three brands of gasoline do not all produce the same mpg. We know that it is possible that we may have made an incorrect decision or inference. From the Fisher table we are much more than 99 percent confident that the three brands of gasoline produce different mpgs. Thus the chance that we made an incorrect inference is much less than one percent. This very small probability is what we called the *p*-value in Chapter 4.

Which brand yields the best mileage figures? The analysis of variance can only tell you that the three brands of gasoline do not all yield the same mpg figures. While you can't tell for sure without doing a further analysis involving confidence intervals, the data suggest that Brand 3 (28 mpg) produces the highest mileage figures.

Why do we need to resort to inductive inference in the analysis of variance? If we wish to draw long-run conclusions from relatively small sample sizes, there's no alternative to the Fisher distribution. It's either inductive inference or an experiment without end—the choice is yours.

In summary there are three steps to doing the analysis of variance.

1. Decompose the total sum of squares into sum of squares–treatment (SSTR) and sum of squares–extraneous factors (SSEF).
2. Convert sum of squares units into variances by dividing by the relevant degrees of freedom.
3. Compute the variance ratio and compare it to the Fisher table.

5:2:4 Running a Valid Experiment

You may not have thought much about how to run a valid study. After all, who has time to worry about such matters when you are facing a mountain of data demanding to be analyzed? But it is important, for if the experiment is improperly conducted, your data is garbage.

Two important design principles are captured by the mnemonic "ROMIN." "RO" means ruling out alternative explanations to the experimental factor effect. "MIN" means minimizing the variance due to extraneous factors. These are discussed in this section and the next one.

Let's return to the brands of gasoline experiment. Forget for the moment how we actually ran the study because we are about to rerun the experiment. Your job will be to detect any flaws in the experiment and correct them.

Since we have three brands of gasoline, let's use three drivers. They will all be driving Ford sedans which will be continually monitored and adjusted when necessary. Let's assign Driver 1 to Brand 1, Driver 2 to Brand 2, and Driver 3 to Brand 3 gasoline. Let's plan to run the five trials for Brand 1 gasoline in the early morning, Brand 2 gasoline during the midday, and Brand 3 gasoline during the late afternoon. Imagine that we again obtain Data Set 1 which generated a variance ratio of over 97. Can we conclude that not all three brands of gasoline yield the same long-run average miles per gallon? The answer is "yes" if we ran a valid experiment and "not sure" if we did not.

In a valid experiment there are no other explanations beyond the three brands of gasoline that could account for the large variance ratio. Ask yourself, are there any other plausible explanations? If you can't think of any, then we probably conducted a valid experiment. Look at the experiment we've just conducted. Can you list two alternative explanations that might account for the large variance ratio?

____

____

____

____

____

____

____

One reason is driver assignment. Each brand of gasoline had its own driver. What may have caused the large variance ratio is not the brands of gasoline, but rather the effect of the three different drivers. We know that drivers can have an effect on miles per gallon—hotrod Harry versus chugalong Charlie. Since each driver drove a car that used only one brand of gasoline, we cannot be sure if the differences in the three brands of gasoline or the three drivers account for the large variance ratio. We have confused the effect of brands with drivers—a seriously flawed experiment.

The second alternative explanation is time of day. Car tires tend to be cool in the morning. By the late afternoon the test track heats up which may cause the tires to expand. This in turn will reduce the friction between the tires and road which may increase the miles per gallon of cars with Brand 3 gasoline. If you find a difference between Brand 1 and Brand 3, you cannot be sure if it is due to the gasoline or the time of day when the test trials were conducted. We have again confused two plausible explanations.

In summary, we cannot rule out two other reasonable explanations for the large variance ratio. We have conducted an invalid experiment and no amount of analysis will change that fact. We must rerun our experiment and minimize the plausibility of the other two alternative explanations.

How would you do it? Are you thinking you'll choose only one driver and one time of day to run the experiment? While this solves the previous problems, it does create other ones. The experiment will now take longer to run and the longer it takes, the greater the likelihood of something else going wrong—Murphy's Law. If we use only one driver, somebody will ask whether it will be possible to generalize our experimental findings beyond our single driver. So let's keep three drivers and plan to run our test cars throughout the day. How can you design the study so as to rule out other plausible explanations beyond the brands of gasoline?

The secret is to ensure that each brand of gasoline is tested by more than one driver throughout the day. If all three drivers test each brand of gasoline and the variance ratio is still large, then driver differences can be ruled out as an alternative explanation. If each brand of gasoline is tested throughout the day, then time of day differences can also be ruled out. If these alternative explanations are ruled out, then the only plausible explanation that could account for a variance ratio of almost 98 is the brands of gasoline.

How do we ensure that drivers are distributed among the three brands of gasoline and that the brands are tested throughout the day?

Randomization!! We assign the drivers to the five runs of each brand randomly. Thus it is unlikely that a single driver will make all five brand runs. We also assign the run times for a brand randomly. It is very unlikely that Brand 1 would be only run in the early morning and Brand 3 would be only run in the late afternoon. Randomization helps us to rule out these two plausible explanations. It also tends to rule out all other possible explanations other than the experimental factor (brands of gasoline). The concept of randomization is a simple tool in experimental design, yet it is powerful and indispensable. Without it, it would be difficult to conduct valid experiments.

Please review the objectives. Can you draw a spread chart? Can you explain the logic of the decomposition in plain English? Can you design a valid experiment? If you can do these three things, you are ready to learn other experimental designs. What are they, when would you use them, and what are the advantages and disadvantages of these designs? That's our next topic.

Exercise Set for 5:2

1. This study assesses the impact of three different merchandising displays on net sales. The dependent variable is sales revenue in tens of thousands of dollars. At what level of confidence can you reject the no long-run difference null hypothesis?

Display 1	Display 2	Display 3
4	6	7
2	5	7
3	4	7

2. If you compute a variance ratio and it is negative, what can you conclude and why?

3. Below are data from an experiment in which three drugs to gain weight were tested on a group of subjects. All the data are in pounds gained during the month. At what level of confidence can you reject the no difference in the long-run weight gain null hypothesis?

Drug 1	Drug 2	Drug 3
6	3	7
6	4	7
8	5	7
8	4	7

4. How would the "RO" principle apply in conducting the previous experiment? Be specific.
5. Two groups of subjects are given the same accounting information. One group is presented the data on a color monitor and the other group on a black and white monitor. The accounting specialist wishes to determine if the presence of color in the presentation of the accounting data improves understanding of the data. She has developed a scale of understandability that goes from zero to five. Zero means that the person doesn't understand the information; a score of five means that the person fully understands what is being presented. The data from the study are shown below.

Color Monitor	Black and White Monitor
5	1
3	1
5	3
4	3
3	2
4	2
4	2

Can you reject the null hypothesis at more than a 90 percent level of confidence? Show all of your work.

6. How do we draw inductive inferences in experimentation and why do we have to do it?

5:3 WHEN TO CONSIDER A ONE FACTOR RANDOMIZED BLOCK DESIGN

By the end of this unit you should be able to:

1. Discuss how the randomized block design accomplishes the "MIN" design principle.

2. Determine using *profile graphs* when a one factor randomized block design is superior to a one factor completely random design.
3. Modify the basic analysis of variance to analyze a one factor randomized block design.

Sir Ronald Fisher invented the randomized block design. The story goes something like this. During World War I, Great Britain was asked to feed the Allied Armies fighting in France. Fisher was in charge of developing a "super fertilizer." After some initial laboratory experiments, he felt it was time for a field test. He subdivided a large tract of land into 12 equal subplots. He then randomly assigned the current/standard fertilizer to six plots and the challenger to the other six plots. After the growing season, he analyzed his results and concluded that the challenger was no better. He found that the variance due to extraneous factors was much larger than he had expected. What had happened?

Fisher believed that one extraneous factor was the primary cause of the larger than expected variance term. Because the test area for his fertilizer experiment had been very large, he reasoned that there could be soil fertility differences in the north-south direction. Remember, England is an island nation. The land closest to the English Channel could be less fertile due to the salt air. Further north, the land was likely to be more fertile.

The challenger (c) and standard (s) fertilizers had each been randomly assigned to six plots throughout the test area—in the north, central, and southern sectors of the test area.

Fisher Study—First Experiment

Northern Sector			
Plot 1(s) 42 bushels	Plot 2(c) 50 bushels	Plot 3(c) 52 bushels	Plot 4(c) 48 bushels
Central Sector			
Plot 5(c) 41 bushels	Plot 6(s) 32 bushels	Plot 7(c) 38 bushels	Plot 8(s) 29 bushels
Southern Sector			
Plot 9(s) 20 bushels	Plot 10(s) 18 bushels	Plot 11(s) 22 bushels	Plot 12(c) 27 bushels

Fisher was correct in that the soil was most fertile in the northern sector. Notice that the standard fertilizer yielded 42 bushels in the northern sector and only between 18 and 22 bushels in the southern sector. Likewise, the challenger yielded between 48 and 52 bushels in the northern sector and only 27 bushels in the southern sector.

Even though it appears that the challenger yielded about ten bushels more than the standard fertilizer, we would not be able to reject the no long-run difference null hypothesis. The tremendous variation due to the different sectors will increase the variance due to extraneous factors and thus make it difficult to reject the null hypothesis.

While the sum of squares due to extraneous factors measures the impact of all extraneous factors, Fisher believed that one extraneous factor—fertility differences in the north-south direction—accounted for a large portion. If he could *extract* (not necessarily eliminate) the effect of the soil fertility differences, this would reduce the sum of squares due to extraneous factors. In turn this might reduce the variance due to extraneous factors, increase the variance ratio, and make it more likely to reject the null hypothesis.

The goal of the "MIN" principle is to *minimize* the variance due to extraneous factors by extracting the impact of one extraneous factor from the sum of squares term.

If Fisher wanted to *eliminate* the effect of soil fertility differences, he could have done all his testing in the northern sector. However, he might never have completed his study in a timely fashion (after all, he couldn't ask the Germans to stop the war to get the test results). Also, he might not have been able to generalize his findings. Will the challenger fertilizer produce the same excellent results in another area of the country? This was the problem facing Fisher. How should he have designed his study so that it would be possible to extract out the impact of soil fertility differences from sum of squares due to extraneous factors, without reducing the external validity of his findings?

In the first study, Fisher had randomly assigned the fertilizers to the individual plots. There was no guarantee that each sector, or block of land (north, central, and south), would have an equal number of standard and challenger fertilizer treatments assigned to it. In fact, there were three plots assigned to the challenger in the northern sector (versus only one for the standard fertilizer) and three plots assigned to the standard fertilizer in the southern sector. He reasoned that if each fertilizer was applied to an equal number of plots within a block of land, he could eliminate the impact of soil fertility on the sum of squares due to extraneous factors.

He reran his experiment in the following way. He assigned the standard and challenger fertilizers to two plots each in the northern block of land. The actual assignment was done using a random numbers table. He repeated this strategy for the central and southern blocks of land. He had developed the randomized block design.

As Fisher defined it, a block is a sector of land which has the following properties:

1. *little* differences in soil fertility *within* a block of land
2. *large* differences in soil fertility *between* blocks of land

But a block need not be a physical land area. What is important is that we expect large differences in the extraneous factor between blocks and little differences within blocks. We also believe that the extraneous factor affects the dependent variable (crop yield in Fisher's study). Later I'll expand on the concept of a block.

Imagine Fisher has rerun his experiment and obtained the following data.

Fisher Study—Second Experiment

Northern Block			
Plot 1(s)	Plot 2(c)	Plot 3(c)	Plot 4(s)
42 bushels	50 bushels	52 bushels	38 bushels
Central Block			
Plot 5(c)	Plot 6(s)	Plot 7(c)	Plot 8(s)
41 bushels	32 bushels	38 bushels	29 bushels
Southern Block			
Plot 9(c)	Plot 10(s)	Plot 11(s)	Plot 12(c)
28 bushels	18 bushels	22 bushels	27 bushels

The major difference between this experiment and the previous one is that both fertilizers were tested twice on each block of land. We have just run a randomized block design and will be able to extract out the impact of differing soil fertility. Let's rearrange the data into our standard format for analysis.

Fisher's Data from the Second Experiment

	Standard Fertilizer	Challenger
	42	50
	38	52
	32	41
	29	38
	18	28
	22	27
Average	30.17	39.33
	Overall Mean 34.75	

The average crop yield in the four plots of land in the northern block is 45.5 bushels (the average of the first two numbers in both columns). In the central block the crop yield is 35 bushels and in the southern block it is 23.75 bushels. The average of those three numbers is 34.75 bushels. Why aren't all the block averages equal to 34.75? After all, each block received both fertilizer treatments. It must be due to soil fertility differences. How can we extract the differences out? While we don't do the formal analysis this way, we could adjust the crop yields of both fertilizers for the effect of the differing soil fertilities.

For the northern block of land, we reduce the actual crop yields by the difference between 45.5 and the overall average of 34.75. Why? Because the additional 10.75 bushels is due to the high soil fertility in the northern block. We do the same for the other two blocks. In that way, the impact of soil fertilizer differences are extracted. Here's how the data would now look.

Data Adjusted for Differing Soil Fertilities

Standard Fertilizer	Challenger	Adjustment Procedure
31.25(N) 27.25(N)	39.25(N) 41.25(N)	Northern block: subtract 10.75
31.75(C) 28.75(C)	40.75(C) 37.75(C)	Central block: subtract .25
29.00(S) 33.00(S)	39.00(S) 38.00(S)	Southern block: add 11.00
30.17	39.33	Sample Average
34.75		Overall Mean

Notice that the spread within the six observations for the standard fertilizer is much smaller than before the adjustment. This is also true for the challenger data. Because of the smaller spreads the sum of squares due to extraneous factors will be dramatically reduced. In the original data from the second experiment the sum of squares–extraneous factors was 980.16 units of variation. It dropped to 33.67 units of variation for the adjusted data. If you have a calculator handy, please verify these two figures. Use the same formula you used to calculate the sum of squares due to extraneous factors for the one factor completely random design.

We don't actually adjust the data to analyze it but that is the effect of extracting out the impact of an important extraneous factor. Having adjusted the data, the variance due to extraneous factors would be small, the variance ratio would be larger, and we would be more likely to find a long-run difference between the fertilizers.

Below are two experiments. Can you explain to your friends why you might want to block on these factors? Remember the blocking factor is not the experimental factor of the study. All we wish to do is remove the impact of the blocking factor from the sum of squares due to extraneous factors term.

Experimental Factor	Potential Blocking Factors
Three types of paint for exterior walls	Painters' experience Orientation of exterior wall (north, south, east, and west walls) Ambient temperature/humidity Exterior surface (concrete, stucco, wood)
Four teaching approaches for a CPA review for accounting personnel	Size of accounting firm (Big 8 versus mom and pop firm) Years since graduation from school Type of degree (MBA versus Masters of Accountancy or Taxation) Time of day (day versus night classes)

The goal of the CPA study is to determine which teaching approach—**experimental factor**—is best. Suppose that we believe that the number of years since an accountant had graduated would have a major impact on the effectiveness of the four teaching methods; there would be a big difference between recent graduates and old-timers' scores on the common final examination. Instead of assigning accountants *totally* at random to the four teaching methods (as in a one factor completely random design), we would block on years since graduation—the **extraneous factor**. Here's the experimental layout.

Block (years since graduation)	**Experimental Factor** Method 1	Method 2	Method 3	Method 4
Less than five years				
5 to 15 years				
More than 15 years				

We would select 12 accountants from a firm. Four accountants would have less than five years experience, four would have between five and fifteen years, and four would have more than fifteen years experience. We would randomly assign the four accountants with less than five years of experience to the four teaching methods, one to each teaching method. We would assign the other eight accountants using the same approach. If there are large differences on the final examination scores between novices, experienced, and senior accountants, we will have accomplished the MIN principle.

5:3:1 Drawing Profiles: A Quick Approach to Determining if Blocking Is Appropriate

Before we begin to torture (analyze) the data, I'll show you how to draw **profiles**. Their purpose is to determine after the experiment has been finished if your choice of a randomized block design was correct. Profiles and spread charts are different. Profiles help you decide whether you should have blocked on a particular *extraneous* factor. Spread charts determine if the *experimental* factor affected the dependent variable; that is, should you reject the null hypothesis.

I've drawn a set of profiles for the data in Fisher's second experiment—his randomized block design study. When drawing profiles, place the dependent variable on the vertical axis and the levels of the experimental factor on the horizontal axis. Here I've used the data from the table on page 196. The standard fertilizer averaged 40 bushels per acre, and the challenger, 51, in the northern block.

Figure 5-3 Profiles for Fisher's Randomized Block Design Data

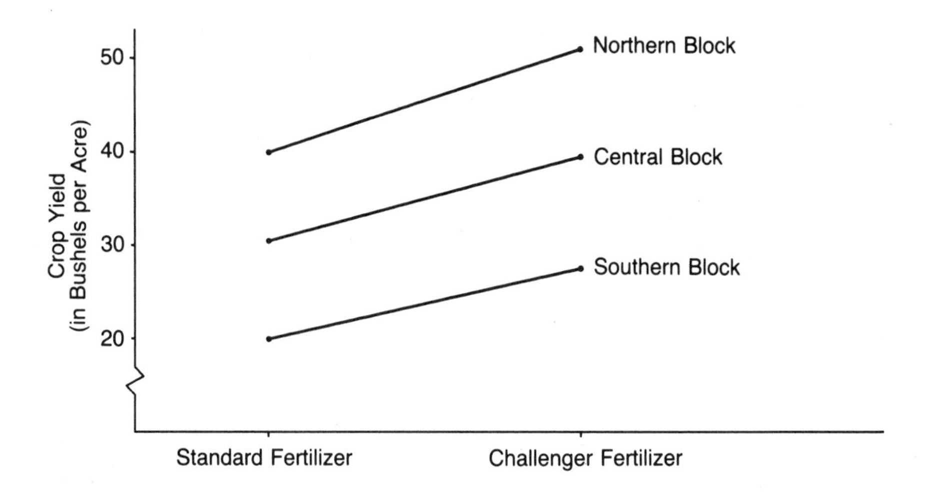

Notice that the profiles are reasonably parallel and spread far apart. The spread is due to the impact of the extraneous factor—soil fertility differences. The greater the impact of the extraneous factor, the greater the spread.

Parallel profiles mean that the apparent improvement in the crop yield of the challenger does not depend on the block of land. The challenger yields about nine to ten bushels more than the standard fertilizer. This is true for *all* the blocks of land—north, central, and southern sectors. Blocking is very appropriate when you expect profiles similar to those in the above figure.

You should block on an extraneous factor when you expect that once the experiment is completed you will get widely spaced and reasonably parallel profiles.

Here's another example. Imagine we have conducted the paint study described earlier, blocked on orientation of the exterior wall, and have plotted the profiles. Blocking on wall orientation requires that we subdivide *each* wall into three equal segments (remember we are testing three paints). Here are two possible sets of results.

Figure 5-4 Two Profiles for Paint Study Data

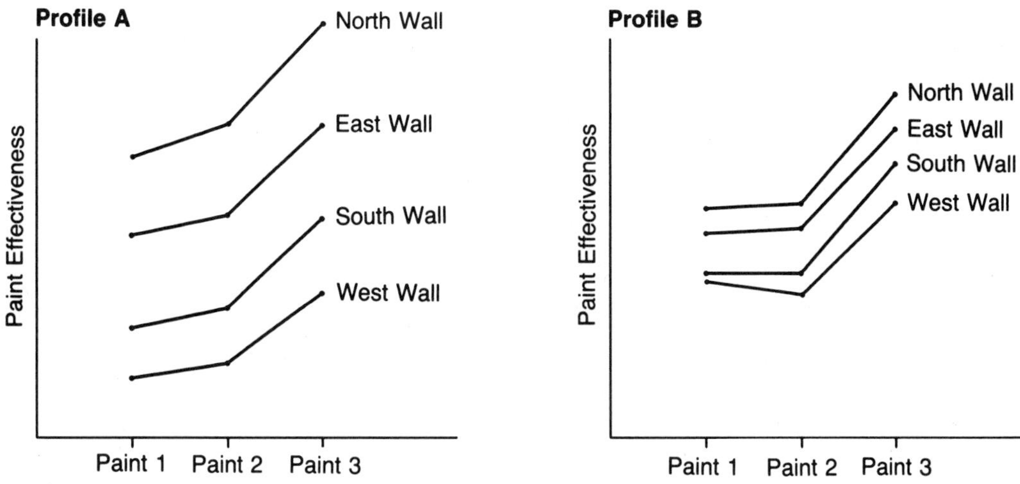

When you chose to block on wall orientation, which profile did you expect to obtain after the experiment was completed? Can you explain why in plain English?

Profile A is what you should expect if blocking were appropriate. In blocking on wall orientation, you believed that:

1. The amount of sun, pollution, and foul weather attack on a single wall is relatively constant.
2. The north, south, east, and west walls are subjected to different amounts of pollution and weather.
3. The differing amounts of pollution etc. are likely to have a great effect on the dependent variable, paint effectiveness.

In short, you expected parallel and widely separated profiles.

Before you analyze the data you should always plot the profiles. If they look like Profile A, the randomized block design was probably a good choice. If they look like Profile B, the randomized block design may not have been a very good choice.

It's time for a summary. Under what conditions would you run a randomized block design? Please develop your own rule of thumb before you read the one below.

When you design an experiment, list as many extraneous factors as you can. Examine your list. If there is one factor that you believe will have a major impact on the dependent variable, you should block on that factor. A major impact means that you would expect widely spaced and parallel profiles.

5:3:2 An Intuitive Sum of Squares Decomposition for a Randomized Block Design

Let's rerun the brand of gasoline study once more. As a one factor completely random design, we used three drivers. We assigned the drivers at random to the three brands of gasoline. This approach rules out alternative explanations. Yet when you suspect that one extraneous factor can have a great effect on the dependent variable you should block on it.

Suppose you suspect that there will be large differences between drivers which could affect the mileage figures for the three brands of gasoline. Further you believe that there are little differences within a driver's performance. If you "burn rubber" one day you are likely to burn rubber every day. While two drivers can have different driving habits, a person's driving habits don't change much from day to day. Thus we will block on drivers (similar to soil fertility and wall orientation).

Here's how the experiment will be run. Each driver (the blocking factor) will test all brands of gasoline (the experimental factor). The run times for the experiment were determined randomly. They are shown on page 203 with the actual test results.

Run time	Driver	Brand of Gasoline	Mileage
8:00 am	Hotrod Harry	Brand 2	13 mpg
8:30 am	Hotrod Harry	Brand 3	7 mpg
9:00 am	Chugalong Charlie	Brand 3	12 mpg
9:30 am	Smooth Sally	Brand 1	19 mpg
10:00 am	Chugalong Charlie	Brand 1	13 mpg
10:30 am	Smooth Sally	Brand 2	21 mpg
11:00 am	Chugalong Charlie	Brand 2	17 mpg
11:30 am	Hotrod Harry	Brand 1	10 mpg
12:00 noon	Smooth Sally	Brand 3	14 mpg

Below are the data for the experiment. The experimental factor is brands of gasoline and the extraneous factor that we chose to block on is individual drivers.

Data Set for Gasoline Study

	Experimental Factor		
Blocking Factor	Brand 1	Brand 2	Brand 3
Chugalong Charlie	13 mpg	17 mpg	12 mpg
Hotrod Harry	10 mpg	13 mpg	7 mpg
Smooth Sally	19 mpg	21 mpg	14 mpg
Average	14 mpg	17 mpg	11 mpg

Even before we begin the formal analysis, let's draw a profile graph and guess if blocking appears to have been appropriate. We'll not draw spread charts (see 5:2) as they are only useful for a one factor completely random design.

Figure 5-5 Profiles for Gasoline Study Data

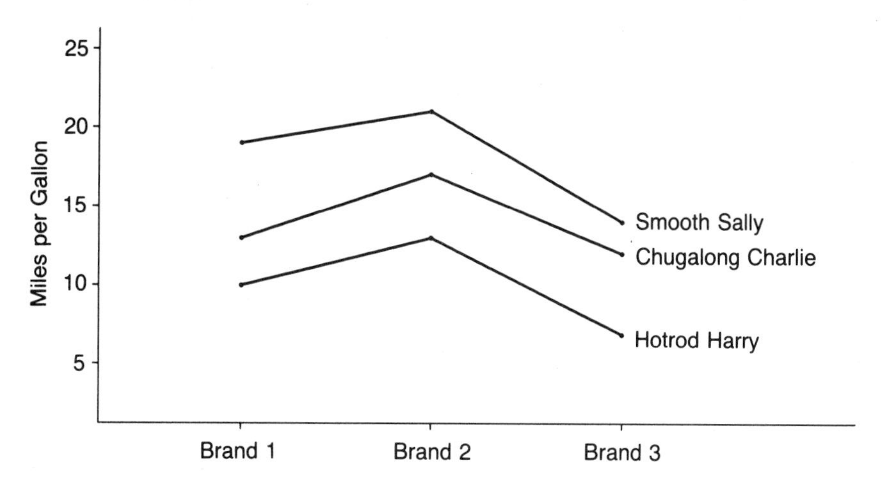

Blocking appears to be appropriate because the profiles are reasonably parallel and spread far apart. We begin the decomposition by computing the sum of squares–total (SST). The overall average is 14 mpg.

$$SST = (13 - 14)^2 + (10 - 14)^2 + \ldots + (7 - 14)^2 + (14 - 14)^2$$
$$= 154 \text{ units of variation}$$

In a one factor completely random design we would decompose SST into two sources of variation: sum of squares–treatment and sum of squares–extraneous factors. For a randomized block design we need to add a third source of variation—sum of squares due to the blocking factor. By blocking we hope that a large portion of what would have been the sum of squares due to extraneous factors for a completely random design will end up in the sum of squares block term. This will reduce the sum of squares due to extraneous factors for the randomized block design. The decomposition for the block design is shown below.

Figure 5-6 Decomposition of Sum of Squares–Total for Randomized Block Design

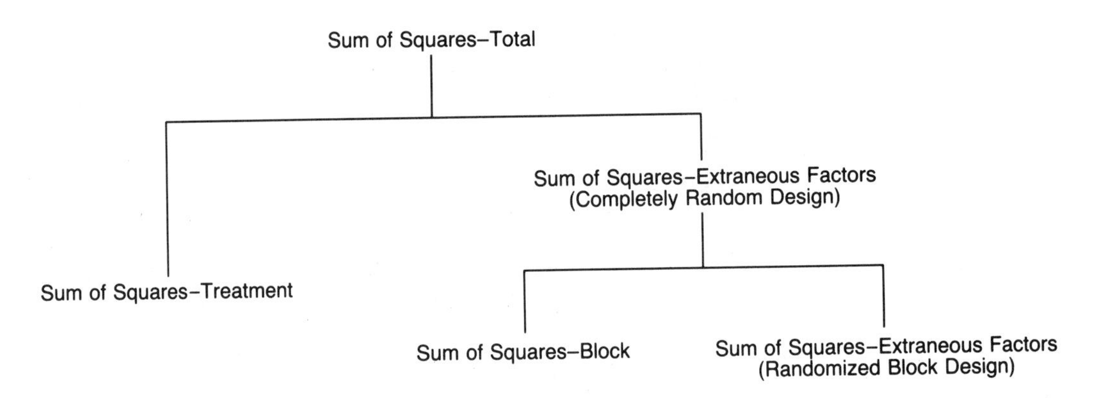

For a randomized block design, there are three sources of variation: the treatment, the blocking factor, and the extraneous factors.

Next we determine the sum of squares–treatment, the variation due to the experimental factor, and the three brands of gasoline. Which three numbers best represent the miles per gallon for each brand?

I hope you concluded the three sample *experimental factor averages*. Don't forget the weighting constant which is the number of observations used to calculate the experimental factor sample averages.

$$\text{SSTR} = 3[(14 - 14)^2 + (17 - 14)^2 + (11 - 14)^2]$$
$$= 54 \text{ units of variation}$$

Now we need to compute the sum of squares blocking factor. Which three numbers best represent the effect of each driver (block)?

These are the three sample block averages which are:

Chugalong Charlie	14 miles per gallon
Hotrod Harry	10 miles per gallon
Smooth Sally	18 miles per gallon

Now what do we do? If these were experimental factor sample averages and you wanted to compute sum of squares–treatment, what would you do? Let's do the same thing to compute sum of squares–blocking factor. How does this look?

$$\text{SSBL} = (14 - 14)^2 + (10 - 14)^2 + (18 - 14)^2$$
$$= 32 \text{ units of variation}$$

It's not quite correct. What's missing?

The weighting constant is not included; that is, the number of observations used to calculate each of the block sample averages. The weighting constant is three. The correct value for the SSBL is

$$\text{SSBL} = 3[(14 - 14)^2 + (10 - 14)^2 + (18 - 14)^2]$$
$$= 96 \text{ units of variation}$$

If the total variation is 154 units and we have accounted for 54 units due to the experimental factor and 96 units due to driver differences (blocking factor), then the remainder is due to extraneous factors. It equals four units of variation.

Sum of squares–total has 9 − 1 or eight degrees of freedom. How many degrees of freedom are associated with the experimental factor (brands of gasoline)? There are three brands; therefore there are two degrees of freedom. For the same reason there are two degrees of freedom associated with the blocking factor. This leaves only four degrees of freedom for the extraneous factors term.

Here's the analysis of variance.

ANOVA Table

Sources of Variation	Sum of Squares	Degrees of Freedom	Variance	Variance Ratio
Block	96	2	—	
Treatment	54	2	27	27
Extraneous	4	4	1	
Total	154	8		

The variance due to the three brands of gasoline is 27 times greater than the variance due to all other extraneous factors; the variance ratio is 27. Compare this value to the Fisher table value for two and four degrees of freedom [F(2, 4, 99 percent) = 18]. We are more than 99 percent confident that the brands of gasoline do affect the miles per gallon. It *appears* that Brand 2 yields the highest mileage (17 mpg).

This example demonstrates how the MIN principle can be accomplished by blocking. Suppose we had run the study as a completely random design and had obtained the same test results (see the data set on page 203). What would the sum of squares due to extraneous factors be if we had not blocked? Please compute it in the following space.

It would equal 100 units of variation. There would also be six degrees of freedom for the extraneous factors term in a completely random design. Thus the variance due to extraneous factors would equal 100 ÷ 6 or 16.67. By blocking we reduced it from 16.67 to 1. That's the MIN principle in action.

5:3:3 What Can Go Wrong When You Choose the Randomized Block Design

The randomized block design is appropriate when you believe that one extraneous factor will account for a large portion of the sum of squares due to extraneous factors. Of course your expectations may be wrong. So what! Is there any disadvantage to choosing the randomized block design even if it turns out that your expectations are incorrect? The answer is *yes*.

What's the problem with running a randomized block design when we should have run a completely random design? If blocking is inappropriate, the sum of squares due to extraneous factors won't be any smaller if we block than if we run a completely random design. But the degrees of freedom for the extraneous factors term is always reduced by blocking. (In the previous example, the degrees of freedom dropped from six for the completely random design to four for the randomized block design.) Thus, instead of *MIN*imizing the variance due to extraneous factors (the sum of squares divided by the degrees of freedom), we actually *increase* it. If the numerator is unchanged and the denominator is smaller, the variance is increased. This makes it harder to reject the no effect null hypothesis.

There is a second design error we could make; that is, running a randomized block design when we should run a multifactor design experiment—our next topic. This design error is illustrated in the following profiles.

Figure 5-7 Profile C for Paint Study Data

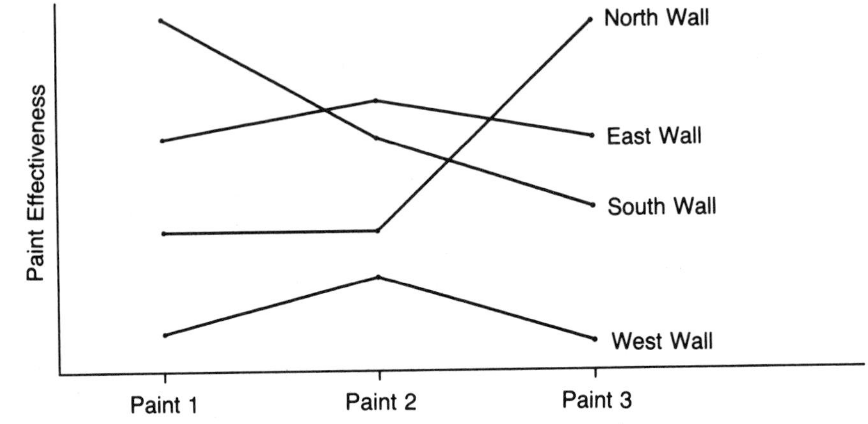

The profiles are spread far apart—that's the good news. This means that the resistance of the paints to the elements varies greatly from one wall to the next. This is an ideal condition for blocking. The bad news is that the profiles are not parallel. This means that the effectiveness of some of the paints varied from one wall to the next. On the north wall Paint 3 was most effective. However, on the west and south walls Paint 3 was least effective. This tells us that the effectiveness of the three paints depended on the location of the wall. This is called an interaction effect. Because we have an **interaction effect** in our study, the level of the dependent variable, paint effectiveness, depends on the interaction between our two factors, the type of paint and the location of the wall. When you anticipate an interaction effect, avoid the randomized block design. I'll discuss why in the next section.

Exercise Set for 5:3

1. This study assesses the impact of four different drugs to reduce blood pressure levels of hypertensive patients. The researcher believes that the effectiveness of each drug will vary widely between patients with different systolic blood pressures. However, she believes that whatever drug is best will be best for all hypertensive patients. She chooses a randomized block design.

   Draw a profile graph and then do the analysis of variance. At what level of confidence can you reject the null hypothesis that there is no long-run difference among the drugs?

Block	Experimental Factor				
	Drug 1	Drug 2	Drug 3	Drug 4	Average
High Systolic	4	7	4	9	6
Very High	7	10	6	9	8
Very, Very High	10	13	11	18	13
Average	7	10	7	12	9

2. How would you select your subjects and assign them to the experimental groups in the drug experiment above? Please be specific so that a third person could actually carry out your instructions.

3. A software firm claims that its new package is faster than its competitors. We wish to test that claim. We believe that there are large differences between microcomputers and we wish to block on them. Each package will be run on each computer. The hypothetical data are shown below and are in seconds. Draw a profile graph and do an analysis of variance. At what level of confidence can you reject the no long-run difference in run time null hypothesis?

Block	Experimental Factor			
	Software 1	Software 2	Software 3	Average
Computer A	8	10	6	8
Computer B	7	9	6	7.33
Computer C	6	8	3	5.67
Average	7	9	5	7

4. We wish to test three additive compounds to see if they improve gas mileage. We decide to block on time of day when the test cars are to be run. The data below are the mpg minus 20 (to make the computations simpler). Plot the profiles before doing the analysis. At what level of confidence would you reject the null hypothesis?

Block	Experimental Factor		
	Compound 1	Compound 2	Compound 3
Morning	4	4	7
Midday	7	6	8
Evening	7	5	6

5:4 WHEN IS THE MULTIFACTOR DESIGN APPROPRIATE?

By the end of this unit you should be able to:

1. Differentiate between the multifactor and the randomized block designs.
2. Draw a set of profiles for a multifactor experiment and determine with 80 percent accuracy whether there is a significant interaction effect.
3. Modify the basic analysis of variance to analyze a multifactor experiment.
4. State in your own words what an interaction is and what its decision making implications are.

You need to differentiate clearly between experimental factors and extraneous factors. It's time for some formal definitions.

Experimental factors are the factors or treatments (brands of gasoline, types of paint) that we wish to test so that we can make timely decisions.

Extraneous factors are all other factors we choose to disregard. We may ignore their impacts and conduct a completely random design. Or we may block on one extraneous factor so as to remove its variation from the sum of squares due to extraneous factors.

A multifactor or factorial design has at least two *experimental* factors and does not use blocking. It relies on randomization to minimize the impact of the remaining extraneous factors. We use a randomized block design when we wish to extract the impact of an *extraneous* factor. We use a multifactor experiment when we:

1. are interested in obtaining information and, in turn, making decisions about two or more experimental factors, or
2. believe that there will be an interaction between all or some of the experimental factors (non-parallel profiles).

Following is the layout for a two factor—four by three—design. Four different teaching methods are to be tested on three different levels of students.

Treatment A: Teaching Approaches

A1 Lecture Method
A2 Discussion Method
A3 Discovery Teaching
A4 Classroom Teaching

Treatment B: Math SAT Score

B1 400–450
B2 500–550
B3 600–650

		Treatment A			
		A1	A2	A3	A4
	B1				
Treatment B	B2				
	B3				

There are 12 (4 × 3) experimental cells or treatment combinations. For example, the subjects in the upper left-hand cell are students with 400 to 450 math SAT scores who are taught by the lecture method.

In designing multifactor experiments you should address the following issues:

1. How many experimental factors or treatments should you include?

You should include those factors that you believe have the greatest impact on the dependent variable. However, the size of the experiment increases very rapidly as you increase the number of experimental treatments.

In this study I chose two treatments—the teaching method and student math ability.

2. How many levels of each treatment should you include?

I recommend that you run at most four levels of a treatment. Beyond that, the size of the experiment rapidly increases and it is difficult to get experimental subjects.

In this study, I chose four levels for the teaching method factor and three levels for the student math ability factor.

3. How many subjects or observations should you have in each experimental cell?

One requirement of a multifactor design is that there must be at least two observations in each experimental cell. The reasons why will be discussed later. You can use as few as two subjects in each cell when you have either a large number of factors or a large number of levels for each factor. It can become difficult finding subjects.

4. How should the subjects be selected and assigned to the experimental cells?

Randomly, of course, for we still must rule out alternative explanations to the experimental factor effects.

Suppose we decide to have two students in each experimental cell. Since there are 12 cells we will need 24 students for our study. First we identify all students with math SAT scores between 400 and 450. Then we randomly select eight students from that population. We use the same procedure to select the other two groups of eight students each. Each group of eight is then randomly subdivided into four subgroups of two students each. We assign each subgroup to one of the four teaching approaches randomly. When the assignment is complete, there will be six students per teaching method.

The multifactor experiment on teaching effectiveness looks similar to the randomized block design, but there is an important difference.

The researcher is interested in both factors in the study. Neither is considered as an extraneous factor. The goal is to determine if teaching approach, math SAT scores, or an interaction between these two factors affect student performance.

5:4:1 Interactions—a Visual Analysis

Imagine we have run the teaching method–student math ability experiment. Look at the two possible profile graphs on the next page.

Does an interaction appear to be present in either set of profiles? If there is an interaction present, what does it mean? If you were in charge of the school, what decisions would you make if you obtained the results shown in each of the two profiles (assuming the findings were verified by the formal statistical analysis)? You may want to review Profile C on page 208 before writing your answer below.

Figure 5-8 Profiles for Teaching Method–Student Math Ability Data

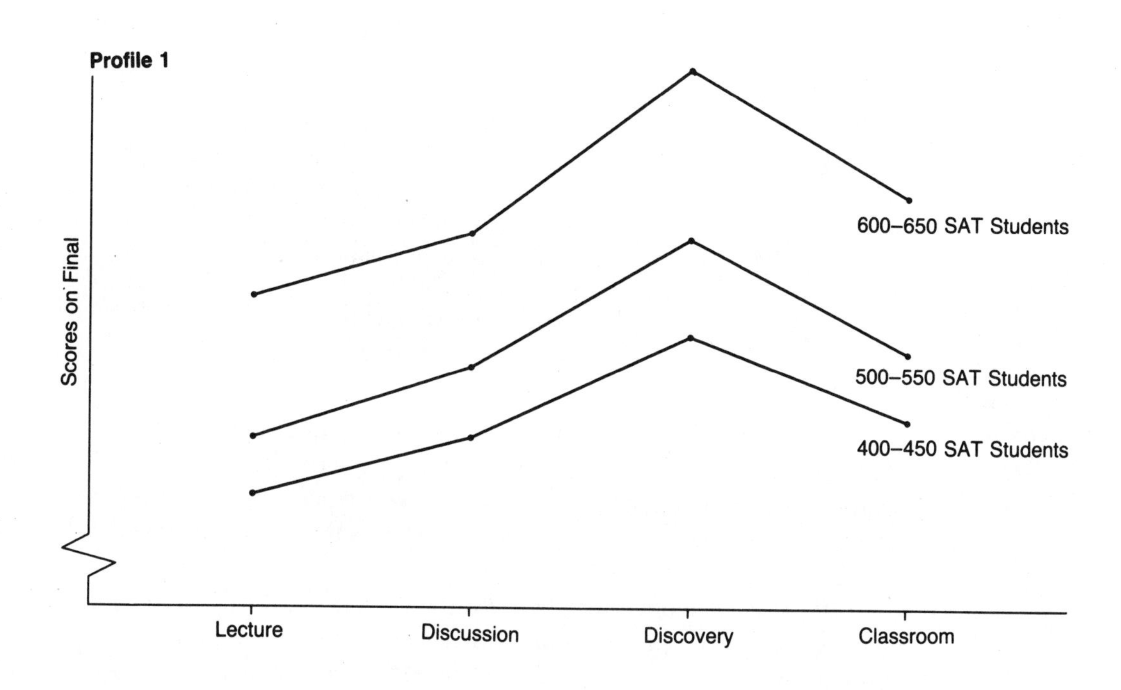

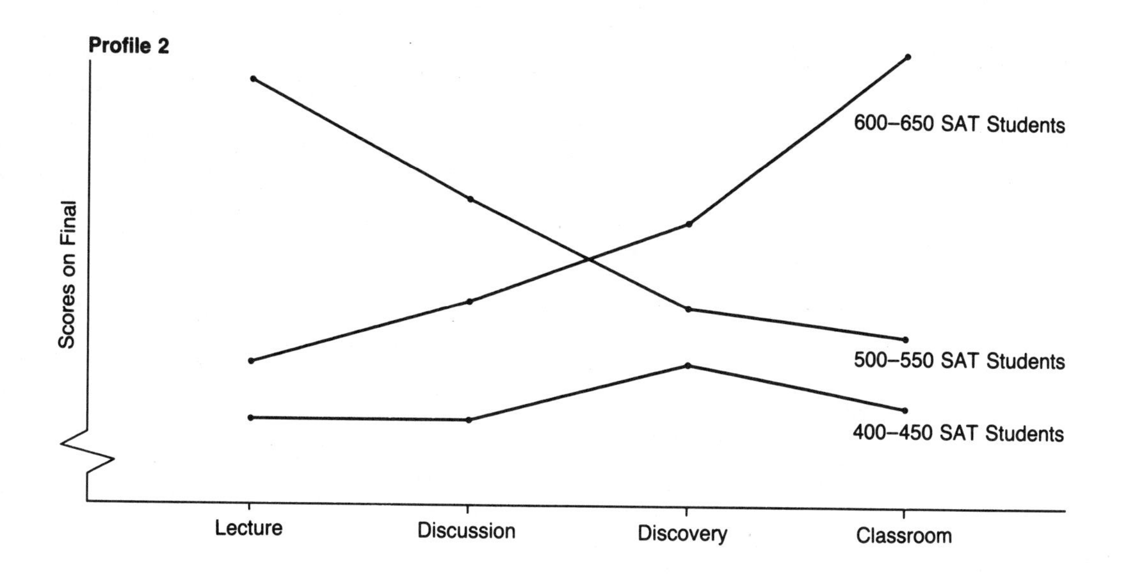

Profile	Intuitive Analysis	Decision Implications
1	No interaction	The profiles are nearly parallel. No matter what the students scored on the math SAT, they did best when taught by the Discovery Method (A3). Implement the Discovery Method for all students in the school.
2	Interaction present	The profiles are not parallel. The best teaching method depends upon the type of student. The brightest math students (B3) did best when the teacher used classroom teaching (A4); the middle group did best when the lecture method was used; and the lowest group did about the same for all four approaches.

When an interaction is present, the best level of one experimental factor *depends* on the level of the other experimental factor. We cannot make general statements such as "one teaching method is best overall." The "best" teaching method depends upon the students' SAT scores.

We still have one piece of unfinished business. Why can't we merely look at the profiles and tell whether there is an interaction present? Why do we have to do a formal statistical analysis and draw inductive inferences?

One reason is that without doing a formal analysis, we cannot even assign a level of confidence that our visual analysis is correct. Beyond that, we are not solely interested in the 24 students in our sample. We wish to use their test results to draw inductive inferences to the entire student body. Without statistical analysis, this cannot be done.

5:4:2 An Intuitive Sum of Squares Decomposition for a Multifactor Design

Below are the data for the teaching approach–student math ability study. Profile 2 was based on this data. Remember that our quick visual analysis indicated that there was *probably* an interaction present. We'll now have a chance to verify it.

	Factor A				
Factor B	Lecture	Discussion	Discovery	Classroom	Average
400–450	58	60	64	59	61.25
	62	60	66	61	
500–550	99	89	80	75	86
	99	91	80	75	
600–650	74	80	85	94	83.75
	76	80	85	96	
Average	78	76.67	76.67	76.67	77

We begin by computing sum of squares–total which represents the total variation in all 24 observations.

$$\text{SST} = (58 - 77)^2 + (62 - 77)^2 + \ldots + (94 - 77)^2 + (96 - 77)^2$$
$$= 4{,}174 \text{ units of variation}$$

What accounts for the 4,174 units of variation? It is due to four effects: the experimental factors A and B, the extraneous factors term, and the interaction effect between teaching approach and the students' SAT scores. Let's compute the first two sources of variation. Don't let the fact that this is a new design throw you. Use your basic definitions and you should have little trouble.

Let's reason through together the computation for the sum of squares–Treatment A (SSTR-A). What four numbers best represent the effect due to the four teaching approaches? Right, it's the four column sample averages. Their average is merely the overall average of 77. Remember that we will have to multiply the squared differences by the number of observations used to calculate each of the four sample averages—the weighting constant. This is the same approach we used in the previous two designs.

$$\text{SSTR-A} = 6[(78 - 77)^2 + (76.67 - 77)^2 + (76.67 - 77)^2 + (76.67 - 77)^2]$$
$$= 7.96 \text{ units of variation}$$

We use the same approach for sum of squares–Treatment B.

$$\text{SSTR-B} = 8[(61.25 - 77)^2 + (86 - 77)^2 + (83.75 - 77)^2]$$
$$= 2{,}997 \text{ units of variation}$$

Next we will calculate the sum of squares due to extraneous factors. Look at the upper left-hand experimental cell where we used the lecture method to teach students with SAT math scores between 400 and 450. Note that the two data values are not the same. Why aren't they since they were both obtained under the same experimental conditions?

The reason that the two numbers differ is the effect of all the extraneous factors that we chose to ignore in the study. These include age of students, motivational level, prior coursework, and others. The cumulative effect of all the extraneous factors may cause the data values within an experimental cell to vary. Does this tell you how to calculate sum of squares due to extraneous factors? Please try it below.

You compute the sum of the squared differences between each cell value and the average for the cell. There are 12 sets of calculations since there are 12 experimental cells. The weighting constant is, of course, one since each data value is a single observation. This is the same procedure we used in the one factor completely random design.

$$
\begin{aligned}
\text{SSEF} = {} & (58 - 60)^2 + (62 - 60)^2 && \leftarrow \text{extraneous variation–Cell 1} \\
& + (60 - 60)^2 + (60 - 60)^2 + && \leftarrow \text{extraneous variation–Cell 2} \\
& \vdots && \vdots \\
& + (85 - 85)^2 + (85 - 85)^2 && \leftarrow \text{extraneous variation–Cell 11} \\
& + (94 - 95)^2 + (96 - 95)^2 && \leftarrow \text{extraneous variation–Cell 12} \\
= {} & 12 \text{ units of variation}
\end{aligned}
$$

The first three factors account for 3,016.96 of the 4,174 units of total variation. The remainder, or 1,157.04 units of variation, must be due to the possible interaction effect.

The total degrees of freedom is the total sample size minus one or 23. The degrees of freedom for Factor A is the number of factor levels minus one or three. The degrees of freedom for Factor B is the number of factor levels minus one or two. Can you determine the degrees of freedom for the extraneous factors term? Remember how we calculated it? Please make an educated guess below and give a brief defense.

As there are two observations per cell, there must be two minus one or one degree of freedom for each cell. There are 12 cells and therefore there are 12 degrees of freedom. Perhaps now you can see why it is necessary to have at least two observations per experimental cell. If we only had one observation there would be no degrees of freedom for the extraneous factors term and we could not compute an ANOVA table.

Thus Treatment A, Treatment B, and the extraneous factors account for 17 of the 23 degrees of freedom. The remaining six degrees of freedom belong to the sum of squares interaction term. Another way to compute the degrees of freedom for the AB interaction term is to multiply the degrees of freedom for Factors A and B ($3 \times 2 = 6$).

Here's the analysis of variance table.

ANOVA for the Multifactor Teaching Experiment

Sources of Variation	Sum of Squares	Degrees of Freedom	Variance	Variance Ratio
Treatment A	7.96	3	2.65	2.65
Treatment B	2,997	2	1,498.5	1,498.5
Interaction	1,157.04	6	192.8	192.8
Extraneous	12	12	1	
Total	4,174	23		

As in all the previous designs, the denominator of the variance ratio is the variance due to extraneous factors. Note that the variance due to the interaction is almost 193 times larger than the variance due to extraneous factors. We now compare that variance ratio against the value from the Fisher table [F(6, 12, 99 percent) = 4.82]. Since our variance ratio is much larger than the Fisher value we are much more than 99 percent confident that there is an interaction effect. This verifies our quick visual analysis of Profile 2.

An interaction tells us that there will be *no* overall best teaching approach; the best approach depends on the students' math SAT scores. Profile 2 suggests (and this too must be subjected to a formal confidence interval analysis for verification) that

1. the brightest students (B3 level) do best when the classroom teaching approach (A4) is used;
2. the mid-level students (B2 level) do best when the lecture approach (A1) is used; and
3. the lowest level students do the same no matter which approach is used.

When you are very confident of an interaction effect (as you are in this study), there is no need to test for the effects due to Factors A or B (the main effects). In other words, you don't have to compute the variance ratios for Factors A or B.

What is the purpose of testing the main effects? Our twin goals are to determine (1) if an experimental factor affects the dependent variable and (2) if it does, what the best level of the factor is. In the education study, the question is, does the teaching approach (or student math abilities) affect student performance on the common examination? If it does, which teaching approach is best? Once we have concluded that the interaction is significant we know that there is *no* best teaching approach; it depends upon the students' math abilities. Therefore there is no need to test for the main effects because the presence of an interaction tells us who should be assigned to the various teaching approaches.

If we weren't confident that there was an interaction, then we must test for the main effects. Had we found that one teaching method was best, it would have been best for *all* students since there was no interaction.

Let's return a final time to the brands of gasoline study. Based upon previously published studies, we decide to include tire pressure as an additional experimental factor. Three brands of gasoline will be tested at two different tire pressure settings. Our study is a three by two multifactor design.

Factor A:		**Factor B:**	
A1	Brand 1	B1	tire pressure at 25 pounds
A2	Brand 2	B2	tire pressure at 35 pounds
A3	Brand 3		

We decide to run three observations per experimental cell. Three identically-equipped cars and two drivers will be selected at random from their respective populations. We'll randomly assign the drivers and cars to the experimental conditions. We will randomly determine the test sequence; that is, which experimental conditions will be tested first, second, etc. Thus it is unlikely that all the test runs using Brand 1 gasoline will be made at the same time each day. By using randomization, we will be able to rule out alternative explanations to a significant factor or interaction effect. Imagine we ran the study and obtained the data shown on the next page.

Multifactor Mileage Study

Factor B	Factor A: Brand 1	Brand 2	Brand 3	Average
25 pounds	18 22 20	24 24 24	29 27 28	24
35 pounds	24 25 23	26 27 28	33 33 33	28
Average	22	25.5	30.5	26

Below is the set of profiles for this experiment. Do you think there is an interaction effect present?

Figure 5-9 Profiles for Brands and Tire Pressure Study Data

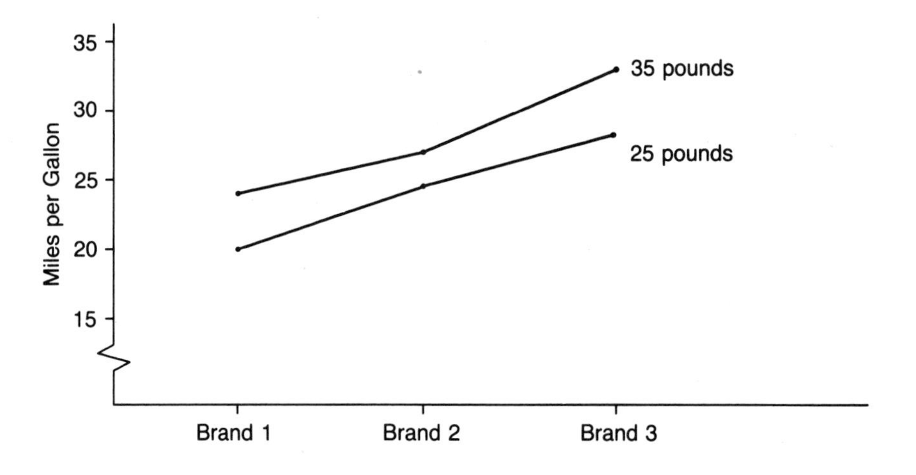

The profiles are nearly parallel. This suggests that there probably isn't a significant interaction effect. Let's do the formal analysis. On the next page I've set up the analysis of variance table but haven't filled in the values. You should be able to complete it in under 30 minutes. Look back to see how we did it before, but please don't look ahead until you've given it a good try.

ANOVA for Mileage Study

Sources of Variation	Sum of Squares	Degrees of Freedom	Variance	Variance Ratio
Treatment A				
Treatment B				
Interaction				
Extraneous				
Total				

ANOVA for Mileage Study

Sources of Variation	Sum of Squares	Degrees of Freedom	Variance	Variance Ratio
Treatment A	219	2	109.5	94.39
Treatment B	72	1	72	62.06
Interaction	3	2	1.5	1.29
Extraneous	14	12	1.16	
Total	308	17		

Again the denominator of the variance ratio is the variance due to extraneous factors. Note that the variance due to the interaction is only 1.29 times larger than the variance due to extraneous factors. We now compare that variance ratio against the value from the Fisher table [F(2, 12, 90 percent) = 2.81]. Since our variance ratio is smaller than the Fisher value we cannot even be 90 percent confident that there is an interaction present. This verifies our profile examination.

We *must* test for the two main effects. We are much more than 99 percent confident that Treatment B, or tire pressure, affects the mpg figures. The variance ratio of 62 is much larger than the Fisher value [F(1, 12, 99 percent) = 9.33]. We are also more than 99 percent confident that Treatment A, or brands of gasoline, affects the mpg figures. The variance ratio of 94 is much larger than the Fisher value [F(2, 12, 99 percent) = 6.93].

Given that there are only two tire pressure settings in our study, we can see from the data that inflating the tires to 35 pounds will improve gas mileage. This will be true no matter which brand of gasoline we eventually choose to use. The data suggest that Brand 3 produces the best gas mileage, although we must do a formal analysis using confidence intervals to verify this.

Before concluding the chapter, please develop three rules for helping you decide when to use the three designs presented in this chapter. Try it before reading on.

Rule 1: If you don't expect any of the literally thousands of extraneous factors to have a large effect on the dependent variable, run a one factor completely random design.

Rule 2: If you expect that one extraneous factor will have a large effect on the dependent variable (widely spaced and parallel profiles), run a one factor randomized block design.

Rule 3: If you expect an interaction effect (non-parallel profiles) or wish to test multiple experimental factors, run a multifactor design.

Exercise Set for 5:4

1. This study assesses the simultaneous impacts of style of leadership and need for independence on office worker productivity. Two styles of leadership and two levels of need for independence will be tested in a laboratory setting to determine their effects on subject productivity. The dependent variable is measured using an index of worker productivity. The data from the experiment are shown below.

   Draw a set of profiles. At what level of confidence can you detect an interaction effect?

Factor A: Style of Leadership	**Factor B: Need for Independence**
A1 Participative	B1 Low Need
A2 Autocratic	B2 High Need

Factor B	**Factor A**		Average
	Participative	Autocratic	
Low Need	10	12	14
	14	20	
High Need	16	8	12
	16	8	
Average	14	12	13

2. In plain English, what are the implications of the significant interaction effect you found in the above leadership study?
3. An experiment is conducted to change attitudes toward drinking and driving. The experiment consists of two factors, each at three

levels. Factor A refers to the information presenter. A1 is a former alcoholic, A2 is a victim of a drunk driver, and A3 is a police officer. Factor B refers to the type of presentation. B1 includes vivid pictures of "driving under the influence" accidents, B2 is a statistical presentation on drunk driving and motor vehicle fatalities, and B3 is the story of one specific accident which took the life of a two-year-old child. There are two subjects per cell.

Two weeks after the experiment each subject's attitude towards starting a chapter of Students Against Drunk Drivers (SADD) is measured. Higher values mean a subject is more in favor of starting a chapter.

	Former Alcoholic	**Victim**	**Police Officer**
Pictures	5	5	2
	3	3	2
Data	2	2	1
	2	2	1
Story	6	9	6
	8	7	6

Draw a set of profiles. Do the analysis of variance. What practical recommendations can be made from the study and why?

4. How would you select the 18 students for the previous study? How would you assign them to the nine experimental conditions? Be specific so that someone else could actually run the experiment.

6. Regression Analysis

6:1 INTRODUCTION

In the early days of World War II, America supplied arms and equipment to England. The convoys transporting the equipment were continually under attack by German submarines. If something were not done, England could not have continued its war effort.

The Navy kept records of the amount of tonnage sunk by German submarines for each convoy. Someone suggested plotting the data. The size of the convoy was plotted on the horizontal axis and the tonnage sunk was plotted on the vertical axis, as I have done below. Each dot in this **scatter diagram** represents a convoy—its size and tonnage sunk. The swarm of dots appears to be horizontal, neither sloping upward nor downward from left to right.

Figure 6-1 Scatter Diagram of Convoys

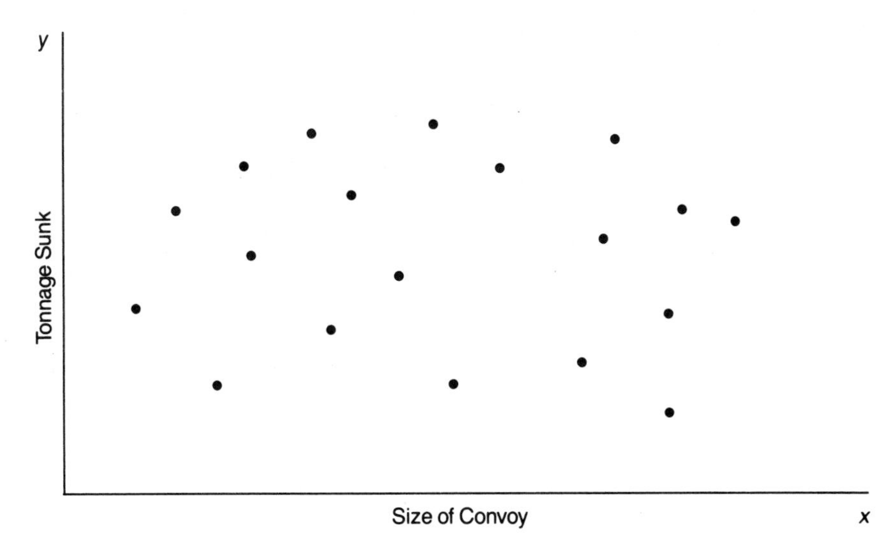

What conclusions can you draw from the scatter diagram? That is, if you had been in charge of organizing convoys, what would you have done to minimize tonnage sunk?

The data suggest that there is *no relationship* between the size of the convoy and the tonnage sunk. Small and large convoys tended to have the same losses. Thus the Navy formulated a new strategy—increase convoy size. This reduced their losses because one large convoy would lose the same tonnage as each of several smaller convoys.

There was no association between size of convoy and tonnage sunk because of the limited number of torpedoes that the German submarines carried. Once all the torpedoes had been fired, the submarines could do no more damage. Eventually the German submarine commanders noticed that the convoys had increased in size and they began to increase the number of submarines in an attack pack. However, by that time the Navy had developed anti-submarine aircraft which protected the convoys. The second Battle of Britain had been won.

In regression studies we are looking for relationships or associations between variables. In this chapter we restrict ourselves to straight line or **linear relationships**.

1. When an increase in one variable tends to be accompanied by an increase in another variable, the variables are positively related.
2. When an increase in one variable tends to be accompanied by a decrease in another variable, the variables are negatively related.
3. When an increase in one variable tends not to be accompanied by either an increase or a decrease in another variable, the variables are not related.

Associations between variables need not be perfect. We know that height and weight of people are positively related. But you know short people who are heavy and tall people who are thin. A positive association between two variables means that, on the average, as one variable increases, the other tends to increase.

Why are associations so important? Whether we find an association or not, we can use that knowledge to make better decisions. In the naval example no relationship was found. Nevertheless, this suggested a strategy to reduce losses. Suppose we run a study to determine if the amount of hands-on training for computer store employees is related to monthly sales. If we find a positive association we will increase the amount of training that each employee receives. This should result in increased sales. Also, we can use the amount of training to predict what a store's sales should be. If the actual sales deviate a lot from the predicted sales, we're going to want to know why. This is the beginning of managerial control.

If there is a strong association between two variables, then you can predict one variable quite accurately by knowing the second one. If there is not a strong association, then you cannot.

In regression studies, there are two types of variables. The **independent variable** is the one that we believe influences the other and is the one that we vary. In the previous chapter, we called the independent variable the experimental factor or treatment. The second variable is called the **dependent variable.** One difference between experimental design and regression analysis is that in regression studies the independent and dependent variables are both quantitative. This means that we can describe them using numbers. Here are some examples.

Independent Variable	Dependent Variable
Hours of marketing training in a department	Sales of the department
Gross income level	Amount of whole life insurance purchased
Amount of tars and nicotine in cigarettes, by brand	Incidence of lung cancer among smokers of those brands
Score on SAT	Grade point average in school

The amount of training is measured in hours and sales are measured in dollars. Gross income and the amount of life insurance are measured in dollars. The amount of tars and nicotine in cigarettes is measured in milligrams while the incidence of lung cancer is measured in cases per thousand in the population. Finally, SAT score and grade point average are measured on quantitative scales.

The statistical methods for studying relationships between variables were developed by Sir Francis Galton in the nineteenth century. He wanted to know how accurately he could predict the height of sons by merely knowing the height of their fathers. He invented regression analysis to answer that question.

You should conduct regression studies when you are looking for relationships or associations between variables in order to make decisions or establish managerial controls. You can exploit associations by making predictions on the dependent variable simply by knowing the value of the independent variable.

6:2 RUNNING REGRESSION STUDIES

By the end of this unit you should be able to:

1. Differentiate between two types of regression studies—experimental and survey.
2. Explain how you use randomization in experimental regression studies.
3. Explain how you use the "pull apart" principle in experimental regression studies.
4. Design an experimental regression study.

You can collect regression data in one of two ways—by formal experiment or by survey. In formal experiments you control the independent variable. This means that you must address the following issues (just as you did in Chapter 5):

1. What should the range of the independent variable be?
2. How many levels of the independent variable should you investigate?
3. How should the levels of the independent variable within the range be spaced?
4. How many observations should you have at each level of the independent variable?

Consider experimental regression studies as one factor, completely randomized experiments in which both the independent and dependent variables are quantitative. The principles of good experimentation you learned in Chapter 5 still apply.

Sometimes it isn't practical to conduct formal experiments. Survey data are obtained without the formal controls of good experimentation. You conduct a survey and use whatever data you obtain from it in your analysis.

Consider a study to detect a relationship between amount of smoking and lung cancer. We could not run a controlled study and assign five different levels of cigarette smoking (the independent variable) to five different groups of people. It would be unethical since we suspect that cigarette smoking is related to cancer. Rather we conduct a survey and ask individuals how many cigarettes they smoke per day. Then we record the incidence of lung cancer among smokers at each level.

Suppose you wish to determine if there is a relationship between the size of a production order and the number of hours it takes to manufacture it. The investigator cannot control the order size (the independent variable). Customer demand for the product determines the order size, not the investigator. Again, you would use survey methods to obtain the regression data.

Now let's focus on experimental regression studies. Here are two examples.

More Training—More Profit? You are in charge of 30 computer stores in the southeast. Several store managers recently requested more technical and hands-on training for their sales personnel. They tell you that more training means more sales. The idea interests you and so you decide to run a regression study to determine if there is a linear association between amount of training and monthly sales. In plain English, does more technical training mean more sales?

The independent variable is the amount of technical training. This means that you believe that training influences sales volume—the dependent variable. Your first decision is what should the range of the independent variable be.

I recommend the **pull apart strategy**. You should select minimum and maximum values for the independent variable that are as far apart as possible. Spread apart the values of the independent variable.

Here's how the pull apart strategy works. Ask yourself, what is the minimum amount of training that is reasonable? You decide that less than four hours is useless. Then ask yourself, what is the maximum amount of training that is practical. You decide that more than 32 hours is too costly. The pull apart strategy says you should make 4 and 32 hours the lower and upper values for the range of the independent variable. I'll discuss why this strategy is important later in the chapter.

Next you must determine the numbers of levels or values of the independent variable. In experimental design the equivalent problem was determining how many levels of the experimental factor we should have in our study. My recommendation is to run as many levels of the independent variable as you are interested in. However, as the number of levels increases, so does the size of the study. This may make it difficult to find subjects. You decide to run eight levels of the independent variable in the training experiment.

It's a good idea to use equal spacing between the levels of the independent variable. Given eight levels and a range from 4 to 32 hours, you should space the training levels at four-hour intervals. Your eight levels in the study are 4, 8, 12, 16, 20, 24, 28, and 32 hours of technical training.

Then you must decide on how many observations you want at each of the eight levels. It's also a good idea to have the same number of observations at each level. You decide to have only one observation per level; this will require eight stores for your study.

Finally, you select the eight stores from the 30 under your control. The stores should be about the same size and have sales personnel with the same number of years of experience. You then assign a store randomly to one of the eight levels of the independent variable. Each of its sales personnel then receives the appropriate amount of technical training. By using randomization and controlling for the store size and years of experience you will be able to rule out alternative explanations if you reject the "no linear association" null hypothesis. Once the store personnel have been trained, you then record their average sales for the following months.

Aerobic Exercise and Your Pulse Rate. Some fitness experts claim that aerobic exercise will reduce your pulse rate and make you live longer. Others say that you don't live longer—with all that exercise, it just seems longer. We'll leave that controversy alone. However, let's determine if there is an association between the amount of aerobic exercise and pulse rate.

Again, the independent variable is the amount of aerobic exercise in hours per week. Amount of exercise is the independent variable because we can control it; we'll test to see if pulse rate changes as we vary the amount of exercise. First we must decide on the range of the independent variable. Using the pull apart strategy, we decide that zero hours will be the lowest value and ten hours per week will be the highest value. Only athletes (and masochists) exercise beyond ten hours per week. We are only interested in drawing inductive inferences to non-athletes.

Next, we decide that we will run 11 levels of the independent variable, spacing them one hour apart. Thus the values of the independent variable are zero, one, two, three, four, five, six, seven, eight, nine, and ten hours of exercise. Finally we select two subjects for each level of the independent variable. After all, it is easy and cheap to find subjects for an aerobic study.

We decide to control the study by selecting only females who are between 25 and 30 years of age and who are all in good health. We select 22 individuals at random from this population. Then we randomly assign each of them to one of the 11 values of the independent variable, making sure that two subjects are assigned to each level. After sixteen weeks of exercising at the assigned level, their pulse rates are recorded upon rising in the morning.

In selecting the values of the independent variable for a regression experiment, you should use the following guidelines:

1. Have at least three levels of the independent variable. This will help you to determine if you can represent the relationship between the two variables with a straight line. For example, suppose that 20 hours of technical training results in the highest monthly sales volume. Sales revenue increases with up to 20 hours of training, but begins to decrease beyond that—a *nonlinear* relationship. You would never detect this by running only two levels (say, 4 and 32 hours). By running a third level you might be able to tell that hours of training and sales volume are nonlinearly related.

2. Have an equal number of observations per level and use equally spaced levels.

3. Be sure to pull apart the levels or values of the independent variable.

Once you complete the experiment, the next step is to analyze the data. You can use the following procedure for both experimental and survey regression studies:

1. Draw a scatter diagram and do a quick visual analysis to determine if it appears that the two variables are linearly related.

2. Determine the best fitting linear line. This is called the least squares regression line.
3. Do an analysis of variance to determine if you should reject the "no linear association" null hypothesis. This is similar to what we did in experimental design.
4. Based upon the analysis, make your decision; for example, should you provide more technical training for your employees. If the two variables are related you can use the regression line to make predictions on the dependent variable.

6:3 A QUICK VISUAL ANALYSIS—THE CIRCLE OR THE ELLIPSE

By the end of this unit you should be able to:

1. Plot scatter diagrams and *estimate* whether or not there is a linear, or straight line, relationship between two variables.
2. Explain the logic to others.
3. Explain why the quick analysis is not a substitute for the analysis of variance.

Before you analyze the data, draw a scatter diagram. Each observation consists of a pair of values—the level of the independent variable and the corresponding value of the dependent value. Represent this pair of values by drawing a point or dot on a graph. Measure the value of the independent variable along the horizontal x-axis, and the value of the dependent variable along the vertical y-axis. Here's the data and the accompanying scatter diagram for the computer store training experiment. Remember that each dot is an x, y pair of values—amount of training and monthly sales.

Store	x (Hours of Training per Employee)	y (Monthly Sales in Thousands of Dollars)
1	4	10
2	8	12
3	12	15
4	16	15
5	20	19
6	24	18
7	28	23
8	32	24

Figure 6-2 Scatter Diagram: Training–Sales Regression Study

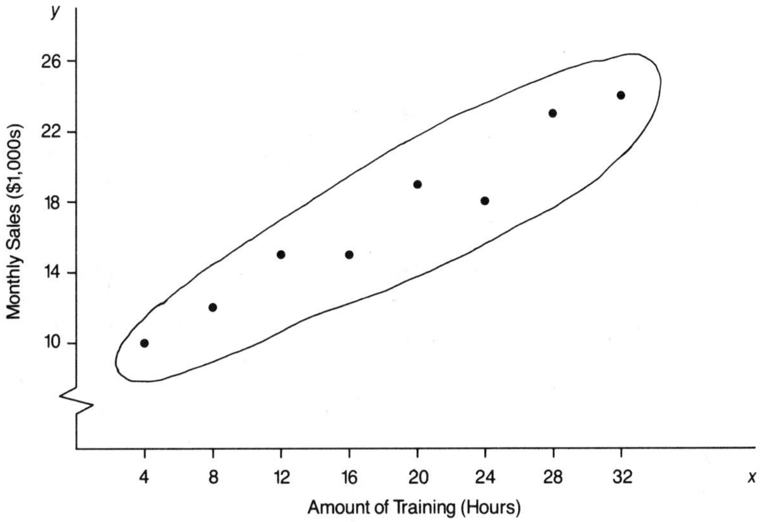

Does it appear that monthly sales, on the average, increase with increasing technical training? Here's a quick test.

If you can enclose all the data points in the scatter diagram with a tight ellipse, then the two variables are probably related. If the ellipse is upward sloping to the right, then the two variables are positively related. If the ellipse is downward sloping to the right, the two variables are negatively related. If it takes a circle to enclose all the data, the two variables are probably not related.

The amount of training and sales volume data can be enclosed by a very tight ellipse that slopes upward to the right. We can conclude that training and sales volume are *probably* positively related.

Why do we have to qualify our conclusion with the word "probably"? This question goes back to the fundamental concept of inductive inference from Chapter 4. Please write your answer in the space below.

Our goal is to draw inductive inferences from the small sample of observations to the population of all 30 computer stores. Without a statistical analysis we cannot even determine what the chances are of being wrong (saying there is a relationship when there isn't or vice versa). We'll use the analysis of variance to make our inductive inferences just as we did in Chapter 5.

Here are three more scatter diagrams for you to examine. Which, if any, suggests an association between the two variables and why?

Figure 6-3 Three Scatter Diagrams

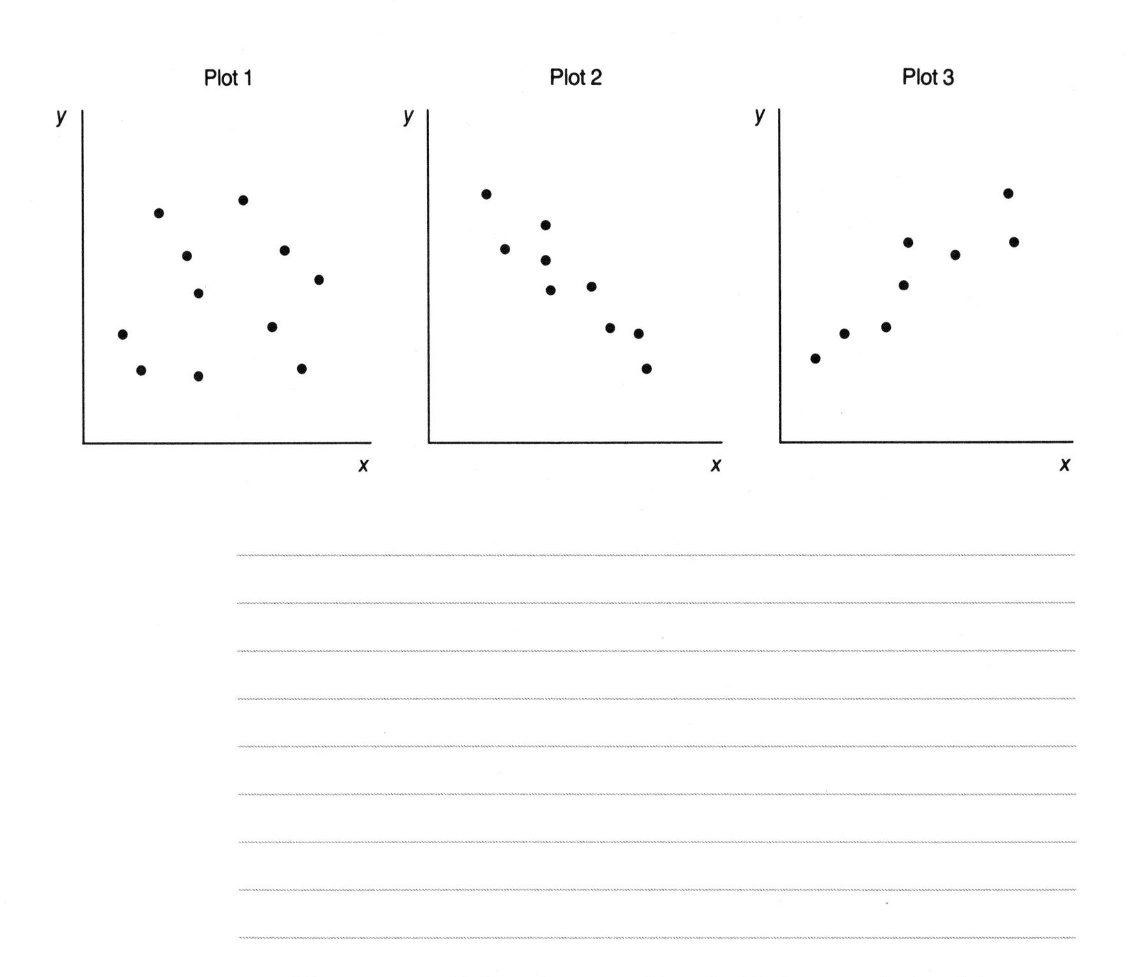

I hope you said that there would probably be no relationship between x and y in Scatter Plot 1. You'll need a circle to enclose all the data points. It appears that in Plot 2 the two variables are negatively related whereas in Plot 3 they appear to be positively related.

Look at Plot 1 where you need a circle to enclose the data. Suppose I ask you to draw a straight line that best represents the swarm of dots. Would the line be horizontal, upward sloping, or downward sloping (left to right)? Please write your answer below.

There is no upward or downward pattern to the swarm of dots; they are scattered all over. As x gets larger, nothing happens to y systematically. Thus your straight line must be horizontal. This means that, subject to statistical verification, the independent variable—x—is not related to y. Also when there is no upward or downward pattern to the dots you need a circle to enclose the data. Thus a circle tends to indicate no relationship between two variables.

In Plot 2 the dots lie in a path from the northwest to the southeast and are tightly clustered. Thus your linear line would be downward sloping. This indicates, subject to statistical verification, that the two variables are negatively related. This means as x gets larger, y gets smaller. When two variables are related, you can enclose their data points in a tight ellipse. The greater the association between the two variables, the tighter the ellipse will be.

The swarm of dots in Plot 3 runs from the southwest to the northeast. We can enclose the data points with a tight ellipse. This indicates, subject to statistical verification, that x and y are positively related.

There is one other reason for plotting the data. Let me illustrate it with the scatter diagram shown in Figure 6-4 on the next page.

Most statistics books focus on looking for **linear**, or **straight line** relationships between variables. The best fitting linear line for all five data points is horizontal (we'll verify this later in an exercise set). This indicates that there is no linear relationship between the two variables. But if you look at the scatter diagram you can see that the two variables are nonlinearly related. If you had stormed ahead with the analysis without drawing a scatter diagram you might have mistaken the finding of no linear relationship for no relationship at all.

Brightman's rule: Plot the data before you start calculating numbers.

Figure 6-4 A Nonlinear Relationship

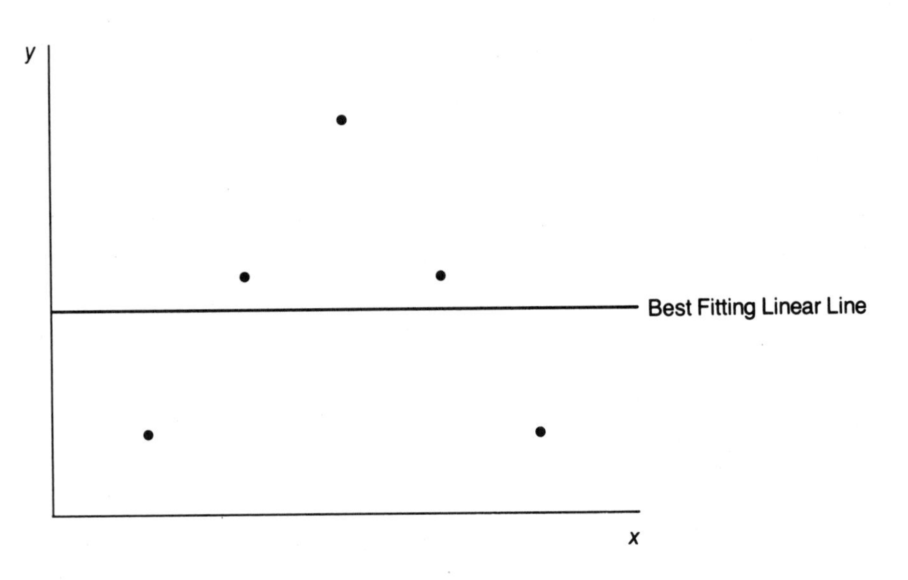

Drawing pictures is useful in computing descriptive statistics for grouped data (Chapter 1), calculating simple probabilities using tree diagrams (Chapter 2), calculating normal probabilities (Chapter 3), constructing confidence intervals (Chapter 4), analyzing experiments using spread charts and profile graphs (Chapter 5), and in regression studies using scatter diagrams.

6:4 DETERMINING THE BEST FITTING STRAIGHT LINE

By the end of this unit you should be able to:

1. Interpret the slope and the intercept.
2. Explain what a deviation is.
3. Explain how you would calculate the best fitting linear line if there were no least squares equations and why this would be extremely time consuming.
4. Solve the least squares equations.

Having plotted the data, our goal is to compute the best fitting linear line to represent the data. Before doing this you need to review some basics about graphing.

Figure 6-5

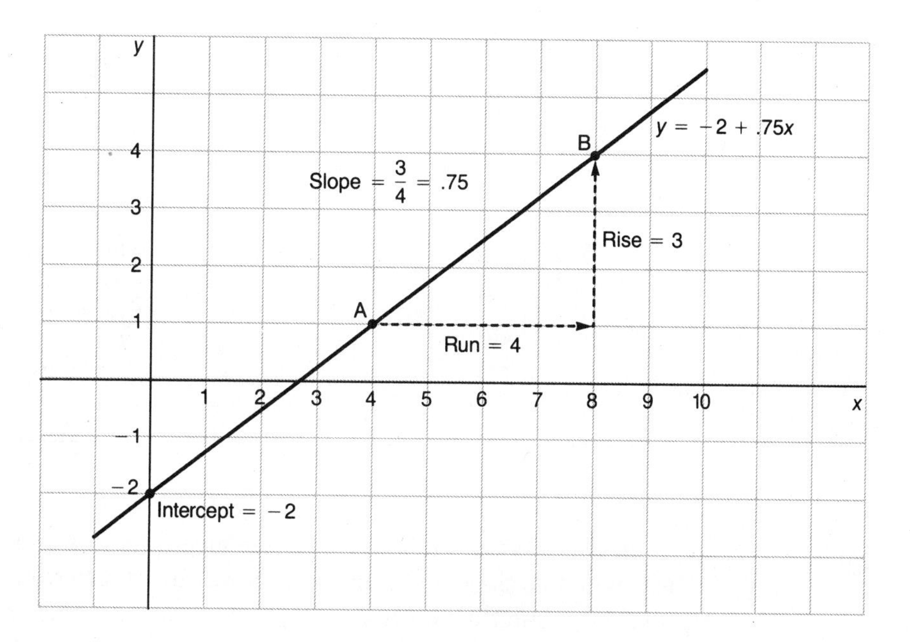

Start at Point A and move up the line to Point B. The distance along the x-axis, four units in this case, is called the **run**. The distance along the y-axis, three units, is called the **rise**. The **slope** is the ratio of the rise to the run, better known as "rise over run."

$$\text{slope} = \frac{\text{rise}}{\text{run}} = \frac{3}{4}$$

The intercept of the line is the value of y where x equals zero. It is −2 in Figure 6-5. The intercept and slope are all you need to define a linear line. The general equation for a straight line is

$$y = a + bx$$

where a is the intercept and b is the slope. The equation of the line in Figure 6-5 is $y = -2 + .75x$. All other points on the line can be determined by simply plugging in the values of x and computing the values of y. For example, when x is 5, y equals $-2 + .75(5)$ or 1.75.

Of course, the slope may be negative as shown in Figure 6-6.

The run is four and the rise (or drop) is −2. Thus the slope is −.5. The intercept is five and the equation for the straight line is $y = 5 - .5x$.

Figure 6-6

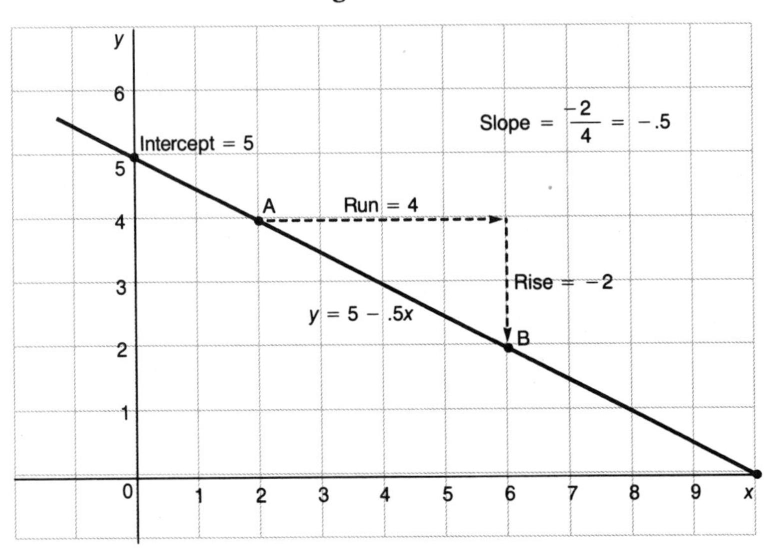

In the previous problems you were given the line and then you determined its slope and intercept. However, in regression studies you must determine the best fitting linear line from the scatter diagram. Before we turn to that, I need to define one more term—the deviation between the actual value of *y* and the value of *y* from the best fitting linear line.

Figure 6-7

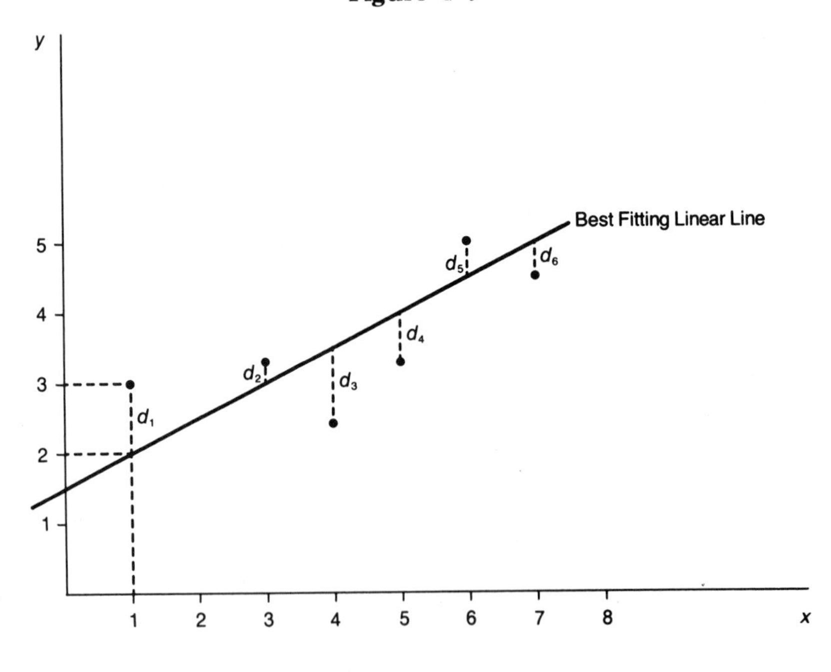

Although we have not yet discussed how to determine a best fitting linear line, let's assume that the line I have drawn is it. Notice that none of the dots are on the line; there is some deviation. When x is one, the actual value of y was three but the value of y from the best fitting linear line was only two. The deviation is $3 - 2$ or one. In general, a **deviation** is just the vertical distance between the actual value of y and the value of y computed from the best fitting linear line. Deviations 1, 2, and 5 are positive and Deviations 3, 4, and 6 are negative.

Now that we have defined the deviation, we need to develop a criterion for a best fitting linear line. Without a criterion how will we know when we have the best fitting linear line? Here it is.

The best fitting linear line is the one that minimizes the sum of the *squared deviations.*

For the data in Figure 6-7, with six observations, the best fitting linear line is the one that minimizes, or results in the lowest possible value of

$$d_1^2 + d_2^2 + d_3^2 + d_4^2 + d_5^2 + d_6^2$$

Picture all the linear lines that could represent the data in Figure 6-7. The best fitting linear line is the one that has the smallest sum of the squared deviations.

Suppose I suggest that the best fitting linear line is the line that minimizes the sum of the deviations (without squaring them). Do you know why this wouldn't work? Here's a hint. Remember deviations can be positive or negative. It will also help if you draw a scatter diagram with only two points and see what happens when you use the sum of the deviations criterion to determine a best fitting linear line. Try it in the space provided below. Don't peek ahead!

Figure 6-8

Figure 6-9

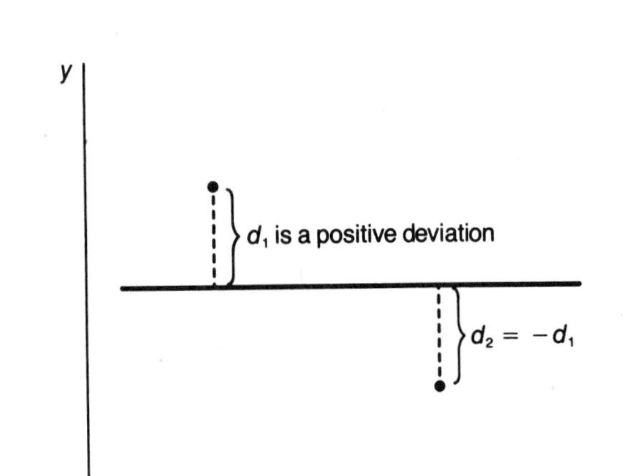

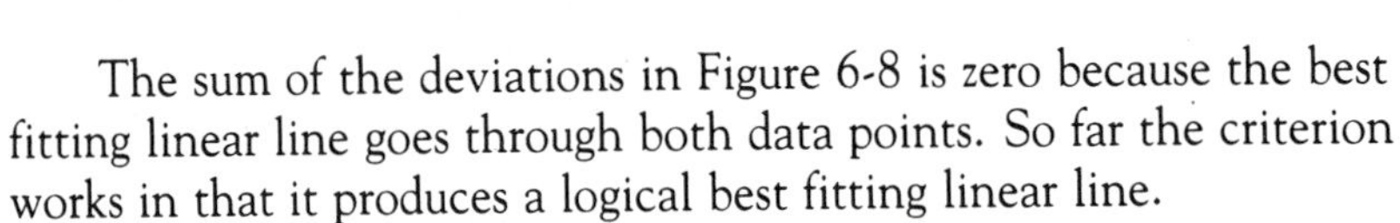

The sum of the deviations in Figure 6-8 is zero because the best fitting linear line goes through both data points. So far the criterion works in that it produces a logical best fitting linear line.

Now for the bad news. The sum of the deviations for the line in Figure 6-9 is also zero. But clearly this is not a very good best fitting linear line because the line is not close to the data points. The problem is clear. Positive deviations cancel out negative deviations and the sum is zero. One way to solve the problem is to square the deviations. That is why we adopted the sum of squared deviations criterion.

Now that we have a criterion, how do we determine the best fitting linear line? There are two ways—trial and error and the correct way. Both ways are demonstrated using the following set of data.

x	y
1	5
2	7
3	6
4	8
5	10

In trial and error you guess at the values of the slope and the intercept. First plot the data. Then pick two data points in the swarm of dots along the major axis of the ellipse. Let's label them A and B. Determine the slope using the rise over run expression. Then determine the intercept by drawing a straight line through the two points and see where it intersects the y-axis. I've done this in Figure 6-10. Using trial and error, my first attempt at a best fitting linear line is $y = 4 + 1x$.

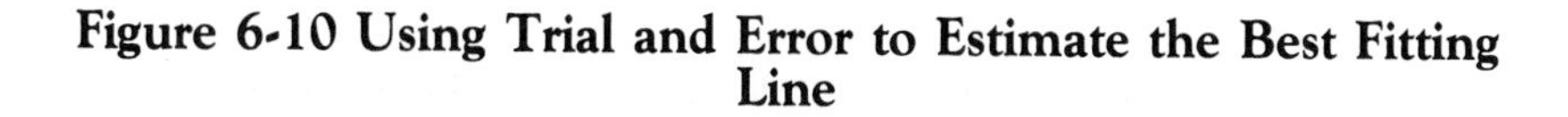

Figure 6-10 Using Trial and Error to Estimate the Best Fitting Line

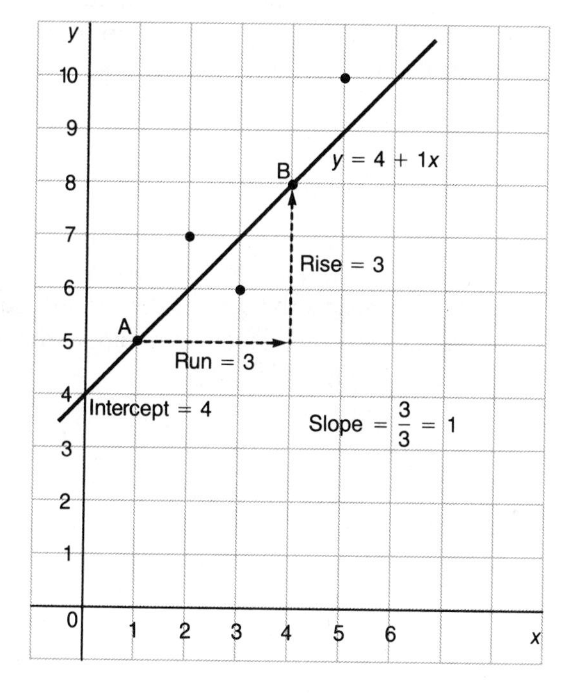

Next compute the sum of the squared deviations as follows:

x	y	y From Trial "Best Fitting" Line	Deviation	Squared Deviation
1	5	4 + 1(1) = 5	0	0
2	7	4 + 1(2) = 6	1	1
3	6	4 + 1(3) = 7	−1	1
4	8	4 + 1(4) = 8	0	0
5	10	4 + 1(5) = 9	1	1
				3

The sum of the squared deviations for the first attempt is 3. Then we would try different values for the slope and the intercept to find a straight line that had a smaller sum of the squared deviations. As you might suspect, trial and error is mostly trial and aggravation. Unless you try all possible values for the intercept and the slope, there is no guarantee that you will ever find those exact values that produce the minimum sum of the squared deviations. There has got to be a faster way.

Through the use of calculus we can develop a set of equations for the slope and intercept that guarantee that we must minimize the sum of the squared deviations. Not surprisingly, these are called the **least squares equations**. Here they are.

$$\text{sample intercept:} \quad a = y\text{-bar} - b(x\text{-bar})$$

$$\text{sample slope:} \quad b = \frac{\Sigma xy - n(x\text{-bar})(y\text{-bar})}{\Sigma(x^2) - n(x\text{-bar})^2}$$

"Σ" is the symbol for "sum of," so Σx means the sum of all the x values, Σxy means the sum of the products of each x times its corresponding y, and so on. Let's apply the least squares equations to the data in Figure 6-10.

$$y\text{-bar} = 7.2$$
$$x\text{-bar} = 3$$
$$n = 5$$
$$\Sigma xy = 109$$
$$\Sigma x^2 = 55$$

$$b = \frac{119 - (5 \times 7.2 \times 3)}{55 - (5 \times 3 \times 3)} = 1.1$$

$$a = 7.2 - (1.1 \times 3) = 3.9$$

$$y = 3.9 + 1.1x$$

Now we can compute the sum of the squared deviations.

x	y	y From Best Fitting Line	Deviation	Squared Deviation
1	5	3.9 + 1.1(1) = 5.0	.0	.00
2	7	3.9 + 1.1(2) = 6.1	.9	.81
3	6	3.9 + 1.1(3) = 7.2	−1.2	1.44
4	8	3.9 + 1.1(4) = 8.3	−.3	.09
5	10	3.9 + 1.1(5) = 9.4	.6	.36
				2.70

The sum of the squared deviations for the best fitting linear line is 2.7. No other straight line will generate a smaller sum of the squared deviations than 2.7. The least squares equations guarantee this.

We are now ready for inductive inferences and the analysis of variance table. Remember, inductive inferences allow you to draw general conclusions about the impact of the independent variable on the dependent variable beyond the experimental units chosen for the study. That's our next topic.

Exercise Set for 6:4

1. Draw a scatter diagram. Use the circle versus ellipse test to see if it appears that the two variables below are linearly related. Then compute the best fitting linear line and determine the sum of the squared deviations.

x	y
1	2
2	4
3	5
4	7
5	12

2. Draw a scatter diagram. Does it appear that the two variables below are linearly related? If not, do they appear to be related at all? Compute the best fitting linear line and determine the sum of the squared deviations.

x	y
1	1
2	3
3	5
4	3
5	1

3. We solved the problem of positive and negative deviations cancelling each other out by squaring the deviations. Can you think of another approach that would solve this problem?

4. Draw a scatter diagram for the next two variables. Does it appear that they are linearly related? Why or why not? Compute the best fitting linear line and determine the sum of the squared deviations.

x	y
1	6
2	4
3	5
4	3
5	1

5. a. Under what conditions can the sum of the squared deviations be negative?

   b. What is the smallest value that the sum of the squared deviations can be?

   c. When the sum of the squared deviations is this lowest possible value, what can you say about the data points and the best fitting linear line?

6:5 INDUCTIVE INFERENCE AND THE ANALYSIS OF VARIANCE

By the end of this unit you should be able to:

1. Explain why we need to resort to inductive inference.
2. Extend the sum of squares decomposition concept to regression studies.
3. Explain the connection between the sum of the squared deviations in regression studies and the sum of squares due to extraneous factors in experimental design.
4. Do the analysis of variance for each problem in the exercise set within 25 minutes using a hand calculator.

Let's return to the computer store training study. Is more training associated with higher monthly sales? The regression data are reproduced below.

Training Hours–Monthly Sales Data

Store	x (Hours of Training)	y (Monthly Sales in $1,000s)
1	4	10
2	8	12
3	12	15
4	16	15
5	20	19
6	24	18
7	28	23
8	32	24

The scatter diagram suggested that the two variables appear to be positively related to one another. We could enclose the data in a tight ellipse that slopes upward to the right. Inductive inference through the analysis of variance allows us to verify our visual analysis, determine a level of confidence for our findings, and generalize the results beyond the eight observations in the study.

6:5:1 Determining the Best Fitting Line

$$y\text{-bar} = 17$$
$$x\text{-bar} = 18$$
$$n = 8$$
$$\Sigma xy = 2{,}780$$
$$\Sigma x^2 = 3{,}264$$

$$b = \frac{2{,}780 - (8 \times 18 \times 17)}{3{,}264 - (8 \times 18 \times 18)} = .494$$

$$a = 17 - .4940 \times 18 = 8.1$$

$$y = 8.1 + .494x$$

You cannot conclude that just because the *sample* slope is positive, there is a long-run positive relationship between the amount of training and monthly sales for *all* 30 computer stores. It's possible that you would have obtained a negative sample slope if you had selected eight other stores. That's why you need the analysis of variance. Without it you cannot assign a level of confidence to your conclusions.

After we determine the sum of the squared deviations we can begin the sum of squares decomposition.

x	y	y From Best Fitting Line	Deviation	Squared Deviation
4	10	8.1 + .494(4) = 10.08	−.08	.0064
8	12	8.1 + .494(8) = 12.05	−.05	.0025
12	15	8.1 + .494(12) = 14.03	.97	.9409
16	15	8.1 + .494(16) = 16.00	−1.00	1.0000
20	19	8.1 + .494(20) = 17.98	1.02	1.0404
24	18	8.1 + .494(24) = 19.96	−1.96	3.8416
28	23	8.1 + .494(28) = 21.93	1.07	1.1449
32	24	8.1 + .494(32) = 23.91	.09	.0081
				7.9848

Figure 6-11 Scatter Diagram with Best Fitting Linear Line

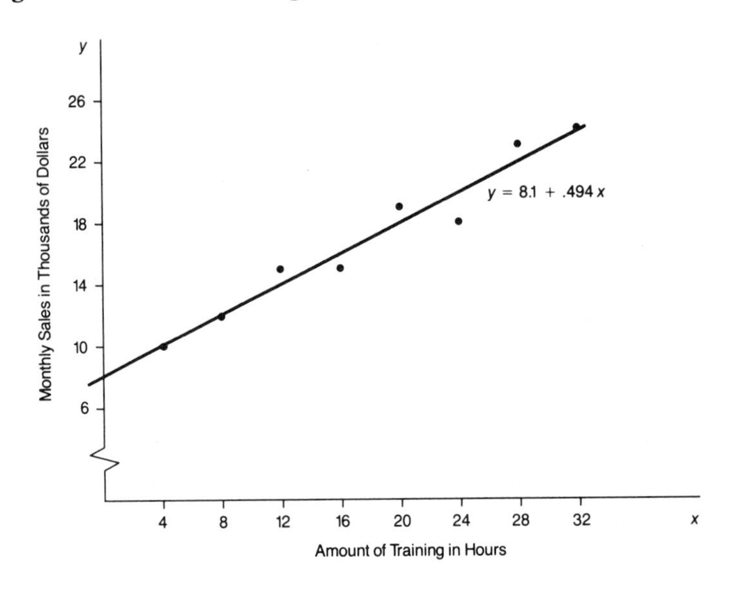

We know that no other straight line will produce a smaller sum of the squared deviations. This is guaranteed by the least squares equations. Now for the sixty-four dollar question. Why isn't the sum of the squared deviations equal to zero? Why don't all the data points lie on the best fitting line? Think about it and jot down your answer below.

The reason it is not zero is due to all other factors or potential independent variables that were omitted from the study. The sum of the squared deviations records the impact of all other factors that affect sales volume. These include seniority of sales force, amount of local competition, the location of the store, etc. If sales were only related to amount of training and there were no measurement errors in recording the data, the sum of the squared deviations would be zero. This never happens in real-world experiments.

In experimental design we had a similar term that recorded the amount of variation due to all other factors omitted in the study. Do you recall what that term was? Please insert it below.

It was sum of squares due to extraneous factors. The sum of the squared deviations is the sum of squares due to extraneous factors in the analysis of variance for a regression study.

6:5:2 The Sum of Squares Decomposition and the Analysis of Variance

The decomposition is similar to a one factor completely random design. We begin with the total variation in the dependent variable, y. How much variation is there in the eight values of y? From Chapter 5, sum of squares–total (SST) is

$$SST = (10 - 17)^2 + (12 - 17)^2 + \ldots + (24 - 17)^2$$
$$= 172 \text{ units of variation.}$$

What causes the 172 units of variation? It is either due to the independent variable or all the other factors we omitted in the study. Thus, sum of squares–total is decomposed into two sources of variation: sum of squares due to the *independent variable (treatment)* and sum of squares due to all *extraneous factors*. Sum of squares due to extraneous factors is the sum of the squared deviations, and is 7.98. Sum of squares due to the treatment is 172 − 7.98 or 164.02.

The total degrees of freedom is still $n - 1$ or 7. Sum of squares due to the treatment measures the impact of the independent variable on y. The impact is represented by the best fitting linear line. To compute it you must know the sample intercept and the sample slope. But once you compute the sample slope, you plug it into the first of the least squares equations to solve for the sample intercept. So although there are two unknowns (a and b), you only have one degree of freedom for the sum of squares due to the treatment. This leaves 7 − 1 or 6 degrees of freedom for sum of squares–extraneous factors.

Let's construct an ANOVA table.

ANOVA Table

Sources of Variation	Sum of Squares	df	Variance	Variance Ratio
Treatment	164.02	1	164.02	123.33
Extraneous	7.98	6	1.33	
Total	172	7		

In regression studies with only one independent variable you will always have $n - 1$ total degrees of freedom. There will always be $2 - 1$ or one degree of freedom for the sum of squares due to the treatment. So there will always be $n - 2$ degrees of freedom, that is, $(n - 1) - 1$, for the sum of squares due to extraneous factors.

6:5:3 The Decision-Making Implications

As in experimental design, we compare the variance ratio to the Fisher table to determine how confident we can be of rejecting the no relationship null hypothesis. Since the variance ratio of 123.32 is much greater than the Fisher value [F(1, 6, 99 percent) = 13.74], we are much more than 99 percent confident that there is a linear relationship between amount of technical training and sales. Since the sample slope is positive, we know that the two variables are *positively* related. You should give additional training to the sales personnel in *all* 30 stores. This should result in increased sales.

Beyond implementing additional training, the best fitting line, or regression line, can be used for setting store sales targets. This is not what the store managers had in mind when they requested the study. Nevertheless, since there is a positive relationship we can use it to set monthly sales target projections as a function of the amount of sales training. You'll see how this is done in the next section. Before that, let's do another example. Below is an example of a survey regression study.

Do people have higher life expectancies in countries that have more doctors? It seems reasonable, but let's investigate. I've used data from 14 randomly selected countries. The independent variable in the study is the population per physician; that is, the number of people for each doctor. The dependent variable is life expectancy at birth in years. The data and a scatter diagram are on page 249.

Regression Study Data: Physicians and Life Expectancy

Country	x (Population per Physician)	y (Life Expectancy at Birth)
Hungary	460	70
France	680	73
Lebanon	1,330	63
Peru	1,800	56
Kenya	5,800	50
Thailand	8,530	58
Ghana	11,220	44
Liberia	11,500	44
Ivory Coast	15,270	44
Senegal	15,360	40
Togo	22,280	41
Guinea	22,380	41
Afghanistan	26,100	35
Mali	33,600	38

Source: Data taken from the 1978 *World Development Report*

Figure 6-12 Scatter Diagram for Life Expectancy Study

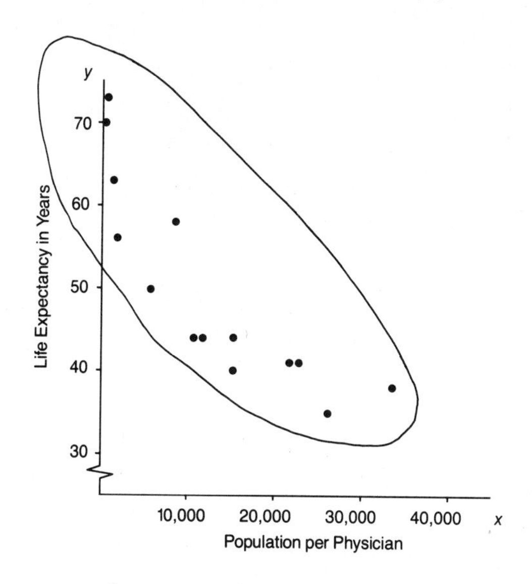

From the scatter diagram it does appear that the variables are negatively related. You can enclose the data points by a tight ellipse that slopes downward to the right. However, before drawing any formal conclusions we need to do an analysis of variance. Unless you have access to a computer don't try to do the actual analysis; it will take too long.

The best fitting straight line is

$$y = 62.431 - .001x.$$

ANOVA Table: Life Expectancy Study

Sources of Variation	Sum of Squares	df	Variance	Variance Ratio
Treatment	1,422.891	1	1,422.891	33.25
Extraneous	513.472	12	42.79	
Total	1,936.363	13		

The variance ratio is 33.25 which is larger than the Fisher value (F[1, 12, 99 percent] = 9.33). We are more than 99 percent confident that as the number of people per physician increases, life expectancy decreases. And we are confident that this is true beyond the 14 countries in our survey. The analysis supports our findings from the quick visual inspection.

Exercise Set for 6:5

1. We wish to determine if hours of aerobic exercise are related to pulse rate. Here's the data.

x (Hours per Week)	y (Pulse Rate)
0	80
2	72
4	65
6	65
8	60
10	58

Draw a scatter diagram, compute the best fitting linear line, and develop an ANOVA table. At what level of confidence can you reject the no linear association null hypothesis?

2. We wish to study the relationship between SAT score and grade point average of college students. Is SAT a good predictor of performance? We run a survey study using only business majors. Thus we will only be able to generalize our findings to business

majors. The hypothetical data are shown below. The independent variable is the combined SAT score divided by 100. The dependent variable is grade point average on a four-point scale.

x (SAT Score ÷ 100)	*y* (Grade Point Average)
7	2.5
8	2.6
8.5	2.9
9.5	2.6
10	2.9
11	3.1
12.5	2.8
13	3.3

Draw a scatter diagram, compute the best fitting linear line, and develop an ANOVA table. At what level of confidence can you reject the no linear association null hypothesis?

3. If sum of squares due to extraneous factors is zero, what can you conclude and why? Remember what the term measures.

4. Do people with higher incomes have more life insurance? We conduct a regression survey study. The independent variable is gross family income and the dependent variable is total amount of "whole life" insurance. The survey data are shown below.

x (Gross Income in $1,000s)	*y* (Insurance Coverage in $1,000s)
13	10
34	20
40	50
55	90
70	125
90	135

Draw a scatter diagram, determine the best fitting linear line, and compute the ANOVA table. At what level of confidence can you reject the no linear association null hypothesis?

5. We wish to determine if there is a relationship between job performance and level of communications skill. Our independent variable is the level of communication as measured by a standardized test. The score on the test can vary from 50 to 100. The dependent variable is job performance as measured by this year's performance rating. The data are on page 252.

x (Communication Score)	y (Job Performance)
50	5
70	7
75	10
85	6
90	5
100	8

Draw a scatter diagram, determine the best fitting linear line, and compute the ANOVA table. At what level of confidence can you reject the no linear association null hypothesis?

6:6 MAKING PREDICTIONS

By the end of this unit you should be able to:

1. Explain why we need to construct prediction intervals when making predictions.
2. Construct prediction intervals.
3. Make managerial decisions based upon the prediction intervals.
4. State three strategies for reducing the width of a prediction interval and explain why you would want to reduce the width.
5. Explain the two reasons why predictions based upon extrapolation are risky business.

Let's return to the amount of training–monthly sales regression study. Here's the equation for the best fitting line.

$$y = 8.1 + .494x$$

x is the amount of training per employee within a store and y is the monthly sales in thousands of dollars.

You are more than 99 percent confident that increased training is related to increased sales volume. You've already decided to give more sales training to the sales force in all 30 stores. This pleases the managers.

They are about to become less happy. Since there is a relationship between the two variables, you can use the regression equation for setting sales targets for the stores. Here's how you do it.

Suppose you would like to predict what the sales should be for a store whose employees received 15 hours of training each. You can use the best fitting line to make the prediction. Substitute $x = 15$ into the equation and determine the expected monthly sales.

$$y = 8.1 + (.494 \times 15) = 15.51 \text{ or } \$15{,}510 \text{ per month}$$

Based upon the best fitting line, a store with 15 hours of training per employee should have sales of $15,510 per month. This is a point estimate. It would be extremely unfair to use only a point estimate as a performance standard. (I'll return to this idea in a moment.) Instead you should set up a prediction interval; this will give each store an interval to shoot for. You expect that their actual sales will fall within the prediction interval most of the time. If their actual sales fall below the interval, the manager better be prepared to explain why. If their actual sales are above the interval, the store's personnel will receive a bonus for doing better than expected. Shortly, I'll show you how to construct a prediction interval.

Why would it be unfair to use a point estimate as a performance target or standard? There are two reasons. First, the best fitting line was computed from a sample of only eight stores. Had the sample been larger, the sample slope and intercept might have been different. This would have produced a different point estimate. So, one reason is that the sample values of the intercept and slope may not be close to their long-run values. Second, we know that other factors beyond amount of training affect the sales volume. Thus a point estimate based only upon training hours is unfair.

Prediction intervals incorporate the uncertainty due to small sample sizes and the impact of all other factors beyond the independent variable. Point estimates do not.

For these two reasons we construct a prediction interval rather than use a point estimate. By the way, how do you know that other factors beyond training hours affect sales volume in the eight stores? Please write your answer below.

I hope you said that the variance due to extraneous factors is 1.33, which captures the impact of all other factors. Since the value is not zero, other factors do have an effect and must be accounted for in the prediction interval.

Here's how you construct a prediction interval. The value of *y* from the best fitting line is called the most likely value (MLV). This is the center of the prediction interval. Similar to confidence intervals from Chapter 4, the expression for a prediction interval is:

$$\text{MLV} \pm t \times \text{estimated standard error of the MLV}$$

The expression for the estimated standard error of the MLV is

$$\sqrt{(\text{Variance-EF})\left[1 + \frac{1}{n} + \frac{(x_p - x\text{-bar})^2}{\Sigma(x - x\text{-bar})^2}\right]}$$

Variance-EF is the variance due to extraneous factors from the ANOVA table. The term "n" is the sample size and the term "x_p" is the value of x for which you wish to make a prediction on y.

We were at least 99 percent confident that training and monthly sales volume are related. Therefore let's construct a 99 percent prediction interval for monthly sales for a store with 15 hours of sales training per employee.

The expression for the prediction interval is:

$$\text{MLV} \pm t \times \text{estimated standard error of MLV}$$

$$\text{MLV} = 8.1 + (.494 \times 15) = 15.51$$

$$t(99 \text{ percent}, 6 \text{ df}) = 3.71$$

$$\text{Variance due to extraneous factors} = 1.33$$

$$n = 8$$

The t value is based upon the degree of confidence and the degrees of freedom for the extraneous factors from the ANOVA table.

$$(x_p - x\text{-bar})^2 = (15 - 18)^2 = 9$$

$$\Sigma(x - x\text{-bar})^2 = (4 - 18)^2 + \ldots + (32 - 18)^2 = 672$$

The estimated standard error of the most likely value is

$$\sqrt{1.33\left(1 + \frac{1}{8} + \frac{9}{672}\right)} = 1.230$$

The 99 percent prediction interval for x equal to 15 hours is

$$15.51 \pm (3.71 \times 1.230)$$

in thousands of dollars, or

$$\$15,510 \pm \$4,563$$

We are 99 percent confident that a store with 15 hours of training per employee should have monthly sales between $10,947 and $20,073. If store sales are within the interval, they are doing what was predicted. If their monthly sales are less than $10,947, they must explain why they are below the lower limit. After all, this should happen by chance only .5 percent of the time. (We expect sales to fall within the interval 99 percent of the time, and of the other 1 percent of the time, half should be above and half below the interval.) There could be extenuating circumstances. A key salesperson was sick or terrible weather kept people from shopping during the test month. If there are no extenuating circumstances, heads will roll. If monthly sales are above $20,073 (again this should happen by chance only .5 percent of the time under normal conditions), they are working above and beyond our expectations. The store's personnel will receive a bonus.

6:6:1 Reducing the Width of Prediction Intervals

As with confidence intervals, we want narrow prediction intervals. A prediction interval that goes from $10,947 to $20,073 is much too wide for control purposes. How can we reduce the width of the prediction interval while keeping the same degree of confidence? There are three strategies for doing this. Two are relatively easy to determine; the third is not. Please review the expression for the prediction interval. What strategies can you use?

You can increase the sample size. This will reduce the second term under the square root radical and thereby reduce the width of the prediction interval. A larger sample size also provides more degrees of freedom for the t value. This reduces the t value and reduces the width of the prediction interval. Of course, increasing the sample size will increase the cost of the study.

A second strategy is to reduce the variance due to extraneous factors. But how can you do this? Take a minute and jot your answer below.

The variance due to extraneous factors measures the impact of all other factors omitted from the regression study. To reduce this variance term you must run studies with more than one independent variable. You have just entered the world of **multiple regression studies** where you have at least two independent variables. A prime motivation for multiple regression is to reduce the width of the prediction interval so that you can make more meaningful predictions.

The third strategy is the least obvious. Look at the denominator of the third term under the radical. If we could make it larger, this would reduce the third term and thereby reduce the width of the prediction interval. How can this be done without increasing the sample size? This is tough so take your time and write your answer below.

I hope you said that if you *pull apart* the values of the independent variable you can increase the denominator. Here's a brief example to illustrate the point.

x Values Bunched Together	x Values Pulled Apart
1	1
2	5
3	9

Let's compute the $\Sigma(x - x\text{-bar})^2$ for each set of data.

$$(1 - 2)^2 + (2 - 2)^2 + (3 - 2)^2 = 2$$

$$(1 - 5)^2 + (5 - 5)^2 + (9 - 5)^2 = 32$$

As you can see, when the values of x, the independent variable, are pulled apart, the denominator of the third term is larger. This reduces the width of the prediction interval.

In the section on running a regression study (Section 6:2) I suggested the pull apart strategy to you. Now you know why. As you spread apart the values or levels of the independent variable, you reduce the width of the prediction interval. The pull apart strategy has one constraint. The values of x you select should be representative of the population to which you want to draw inferences. If you are interested in non-athletes in the pulse rate study, don't select individuals who exercise 30 hours a week. Only long distance runners exercise that long.

There are three ways to reduce the width of a prediction interval. Increase the sample size, include multiple independent variables in the study, and pull apart the values of each of the independent variables in an experimental regression study.

6:6:2 When Predictions Are Risky Business

Let's return once again to the training–monthly sales regression study. Since we are more than 99 percent confident that the two variables were related, we can use the best fitting straight line to make predictions on y for values of x. There are, however, some dangers in making predictions.

In the original study, we varied the independent variable from 4 to 32 hours of training. This is called the *domain* of x. We can use the best fitting line to make predictions within the domain of x. But can we use the best fitting linear line to make prediction outside the domain—for values of x less than 4 or greater than 32 hours? Can we make extrapolating predictions? Yes, but there are two risks. Try to identify the two risks before reading on. One is a logical risk and the other is a statistical risk (look at the expression for the prediction interval).

The logical risk. One problem is that we don't know if the relationship we found within the domain will hold outside that range. Even though we found a positive relationship, we know that at some point increased training will not produce more sales. However, we don't know where the break point will occur. Common sense tells us that you can probably use the regression line for values of x just outside the domain, but it is risky business. The further away from the domain you go, the more risky are the predictions.

The best strategy is to incorporate the range of x values in the original study for which you wish to make predictions. If you want to make predictions for values of x between 4 and 40 hours of training, be sure to include these values in the original study. Then you won't need to make extrapolating predictions. If it turns out that beyond 32 hours the positive relationship doesn't hold (see the scatter diagram below), your scatter diagram should show this to you. You could then develop two best fitting linear lines; one for values of x between 4 and 28 hours and the other between 32 and 40 hours.

The statistical risk. Look at the numerator of the third term under the radical—$(x_p - x\text{-bar})^2$. As x_p diverges from x-bar (which is true of extrapolating predictions) the numerator increases which will widen the prediction interval. Thus, even if the regression line is still valid outside the domain of x, the prediction interval may be so wide as to be almost meaningless.

Figure 6-13

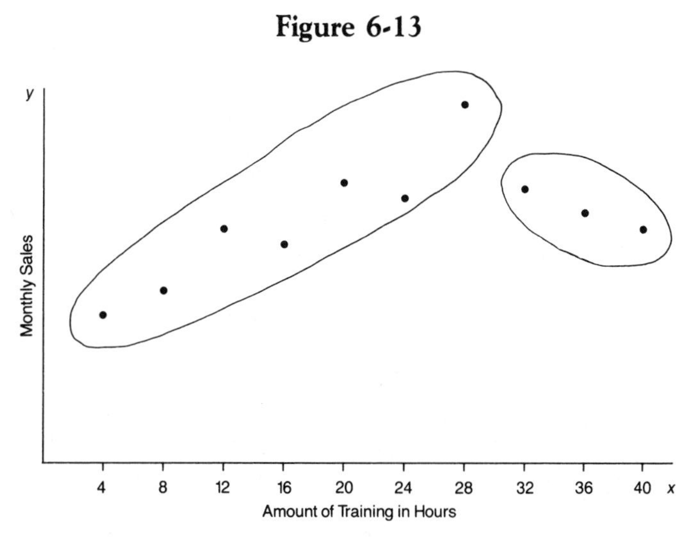

Be cautious when you use the regression line for making predictions outside the domain of x. Do you believe that the regression line is still valid? Will the width of the prediction interval allow you to make meaningful predictions? If the answer to both questions is yes, use it.

Exercise Set for 6:6

1. Construct a 99 percent prediction interval on y for $x = 7$ hours of exercise. Use the best fitting linear line from the Exercise Set for 6:5, Problem 1.
2. How will the three strategies for reducing the width of the prediction interval affect the cost of your study? Why? Can any of them be implemented without increasing costs?
3. Construct a 95 percent prediction interval on y, the amount of whole life insurance, for a person with an income of $80,000. Use the best fitting linear line from the Exercise Set for 6:5, Problem 4.
4. Return to Problem 1 in the Exercise Set for 6:5. First draw the best fitting linear line accurately on graph paper. Construct three 90 percent prediction intervals at $x = 0$ hours, 5 hours, and 10 hours. Plot the upper and lower limits for each prediction interval around the best fitting linear line at the three values of x. Draw a smooth curve through the three upper limits and another smooth curve through the three lower limits. Are the two smooth curves linear or are they curved? What does that tell us in plain English?

6:7 MULTIPLE REGRESSION STUDIES: MANY INDEPENDENT VARIABLES

By the end of this unit you should be able to:

1. Conduct a multiple regression experiment.
2. Explain what multicollinearity is.
3. Assess the degree of multicollinearity.
4. Describe how to minimize multicollinearity.

We use multiple regression when we believe that more than one independent variable affects the dependent variable. If we are right, we may be able to reduce the width of our prediction intervals and thus make more meaningful predictions. There are two types of multiple regression studies—experiments and surveys. We'll focus on the former.

6:7:1 Conducting Multiple Regression Experiments

Earlier we determined that more technical training leads to more sales. Suppose that we decide to add an additional independent variable to our study—the commission rate for the employees within a store. We believe that both more training and higher commission rates lead to more sales. We decide to run three levels of each independent variable. The levels of the technical training independent variable will be 4, 18, and 32 hours. The levels of the commission rate independent variable will be 3, 9, and 15 percent. For reasons that will become clear later we decide to test all nine possible combinations of the two independent variables. The nine combinations are shown below in the 3 × 3 table.

Commission Rate	Hours of Training		
	4 hours	18 hours	32 hours
3 percent			
9 percent			
15 percent			

We select nine stores at random for our study. We randomly assign each of the nine stores to one of the nine treatment combinations.

The upper-left cell represents a store whose sales personnel get 4 hours of training per employee and a 3 percent commission rate. The lower-right cell represents a store whose sales force get 32 hours of training per employee and a 15 percent commission rate.

Imagine the experiment has been completed. Below are the data.

Data for the Multiple Regression Experiment

Store	x_1(Training Hours)	x_2(Commission Rate)	y(Sales in $1,000s)
1	4	3	12
2	18	3	18
3	32	3	20
4	4	9	16
5	18	9	23
6	32	9	24
7	4	15	20
8	18	15	24
9	32	15	28

I ran the data through a multiple regression computer package and obtained the following equation for the best fitting linear *plane*. (You don't do multiple regression by hand unless you're looking for a way to kill a weekend.)

$$y = 9.91 + .285x_1 + .611x_2$$

where y is the monthly sales volume in thousands of dollars, x_1 is the amount of technical training per employee, and x_2 is the commission rate.

We expected that increasing hours of training and commission rates would lead to more sales. It appears that we were correct as the two sample slopes are positive. Nevertheless, we will need to compute an analysis of variance so that we can assign a level of confidence to our findings. Here is the ANOVA table, which was also obtained from the multiple regression computer package.

ANOVA Table

Sources of Variation	Sum of Squares	df	Variance	Variance Ratio
Treatments	176.667	2	88.333	55.451
Extraneous	9.556	6	1.593	
Total	186.223	8		

The variance ratio of 55.451 is much larger than the Fisher value [F(2, 6, 99 percent) = 10.92]. We are much more than 99 percent confident that the two independent variables are related to sales performance of the computer stores. We could now make predictions on *y* using the two independent variables. Most computer packages offer the option of constructing prediction intervals.

6:7:2 Multicollinearity—A Potential Problem with Survey Data

The reason for testing all nine combinations of the two independent variables was to avoid a problem called multicollinearity. The problem does not arise in a well-designed experiment, but you have to be aware of it to design the experiment properly. And it can be a serious problem in multiple regression *survey* data. So we need to discuss multicollinearity—what it is, how you know if you have it, and what you can do to minimize it. Think of multicollinearity as a hurdle. If you don't overcome it, you should not use the output of your multiple regression study.

What is multicollinearity? In regression studies you hope that your independent variables are related to the dependent variable. When the independent variables are related to one another, you have **multicollinearity**.

Severe multicollinearity exists when the independent variables are strongly related to each other. As the values of one independent variable increase do the values of the other independent variable either tend to increase or decrease? If they do, you have multicollinearity.

As there are no formal statistical tests for multicollinearity, I recommend that you draw scatter diagrams for all the pairs of the *independent variables*. You can use the circle verses the ellipse quick informal test. If you can enclose the scatter diagram by a tight ellipse, you have severe multicollinearity; if you need a circle to enclose the data, you don't.

Let's suppose that instead of conducting an experiment we took a survey of ten computer stores to see if training and commission rates were related to sales. Here's the scatter diagram for the two independent variables in our survey.

Figure 6-14 Scatter Diagram for Two Independent Variables

x_1

Amount of Training in Hours

Commission Rate x_2

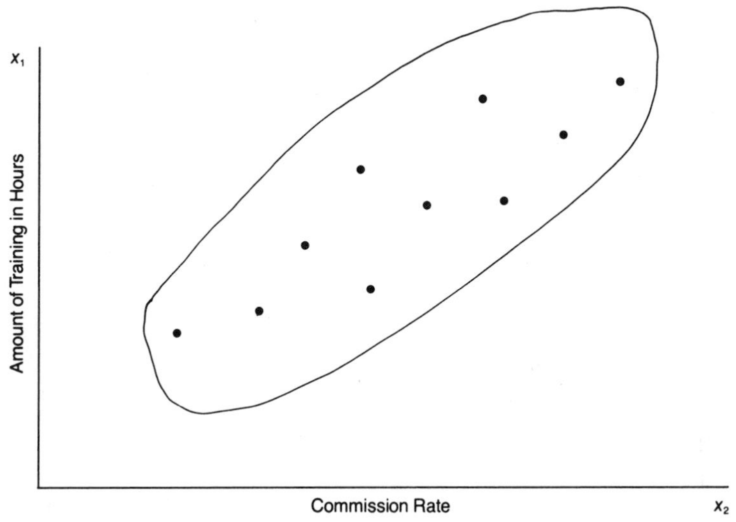

Does there appear to be either a positive or negative relationship between the two independent variables? More specifically, as the number of hours of training increase do the commission rates tend to either increase or decrease? From the diagram, you can see that the two independent variables appear to be positively related. We need an ellipse to enclose the data. We have a multicollinearity problem.

Multicollinearity is *not* an all or nothing proposition. If the independent variables are only slightly related (a wide ellipse encloses the data) you can use the results of the study. But if the independent variables are highly related (a very tight ellipse), you only have two choices. You can either consider a remedial strategy (I'll discuss one below) or forget about using regression analysis.

Why severe multicollinearity is bad. When there is severe multicollinearity, the values of the sample intercept and slopes are not meaningful. For example, if amount of training and commission rate are highly related to one another, how could we predict what would happen to sales (the dependent variable) if only one variable—hours of training—were increased? The problem is that when two independent variables are related, as you vary one the other varies also. Thus you can never determine what the impact of each independent variable is.

In plain English, when you have severe multicollinearity, the sample results are simply not meaningful.

Overcoming multicollinearity. In multiple regression experiments, the best approach is to avoid multicollinearity by designing it out of your experiment. This is the idea behind running all possible treatment combinations. For example, we ran all nine treatment combinations from the 3 × 3 table to avoid multicollinearity in our training and commission rate experiment. Please verify that we have no multicollinearity by drawing a scatter diagram for the two independent variables.

You need a circle to enclose all the points. This is, of course, because we designed the experiment that way. However, in surveys, the analyst does not have full control over the values of the independent variables. He or she must use whatever data or values of the independent variables are obtained in the survey.

One remedial approach is to eliminate all the highly related independent variables from the study. Suppose you have five independent variables, two of which are highly related. You could drop the two variables out of your study. This will eliminate the multicollinearity. Unfortunately, you tend to "throw the baby out with the bath water." If the discarded independent variables were related to y, you may not be able to reject the no linear relationship null hypothesis.

In most multiple regression studies the independent variables will exhibit some degree of relationship. You should only become concerned when the independent variables are highly related to one another.

6:8 ASSOCIATION AND CAUSE AND EFFECT

Regression analysis looks for relationships or associations. What does an association mean? Is association the same as cause and effect? Earlier we found a strong relationship between training hours and monthly sales. Can we say that increasing training causes increased monthly sales? Before answering that question, consider the following study. We sample cities throughout the United States and record the number of clergy and the number of alcoholics. We find a strong positive association between the two variables. Can we conclude that the clergy cause people to drink?

A **causal relationship** exists when changes in one variable cause changes in another variable, *holding all other factors constant.* The case for causation is strengthened when:

1. The association between the two variables is found in many studies done by different investigators.
2. You can provide a reasonable argument as to how changes in one variable could cause changes in the other.
3. There are no other plausible factors that explain the connection between the two variables.

In the sales training study, you can make a reasonable argument that additional hours of training was a cause of the increased sales. After all, the purpose of training is to improve product knowledge which should produce more sales. Are there any other plausible factors that might explain the association between training and sales? For example, if stores that received more training had more senior sales forces or carried higher price merchandise, then these two factors might be reasonable alternative explanations to the association. If these could be ruled out, then the argument that more training hours causes more sales is strengthened.

In the clergy example there are no reasonable explanations to the association. There is also an obvious third variable that affects both the number of clergy and the number of alcoholics—population of the city. As cities increase in population, you have more clergy and more alcoholics, but one does not cause the other.

For many years we knew that there was a strong association between amount of smoking and lung cancer. Can we say that smoking is a cause of lung cancer? First, this association has been found by many investigators throughout the world. Second, investigators, in well-controlled experimental studies, have found that substances in cigarettes can produce cancers in animals. So there is a plausible explanation. Finally, the differences in life styles between smokers and nonsmokers (other than smoking) have not been found to be related to the incidence of lung cancer. It appears that smoking is a (not *the*) cause of lung cancer.

Regression focuses on detecting associations. Assessing causation is beyond statistics. Causation requires association and explanation.

Solutions to the Exercise Sets

Exercise Set Solutions for 1:3

1. The three means are 4, 23.75, and 51,094.2. The three medians are 5, 3.5, and 789. In each case, half the numbers are above and half are below the median value.
2. The large number of older students will shift the mean to the right in the histogram. This is because the mean is the balance point of the histogram. However, most of the students will still be between 18 and 22 years of age, therefore the median will not be as affected as the mean by the addition of older students.
3. The mean is a measure of central tendency, not a measure of spread. The mean depth of the water could be two feet and yet the river could be over six feet deep at spots. That is what he forgot.
4. $\text{mean} = [(47.5 \times 1) + (52.5 \times 2) + (57.5 \times 6) + (62.5 \times 2) + (65.5 \times 1)] \div 12$

   $= 57.5$

   Given 12 observations, the median is halfway between Observations 6 and 7. These two observations lie in the third class. Subdivide the interval from 55 to 60 into six subclasses. Halfway between the sixth and seventh observation is 57.5. Look at Figure 1:3.4.

Figure 1:3.4

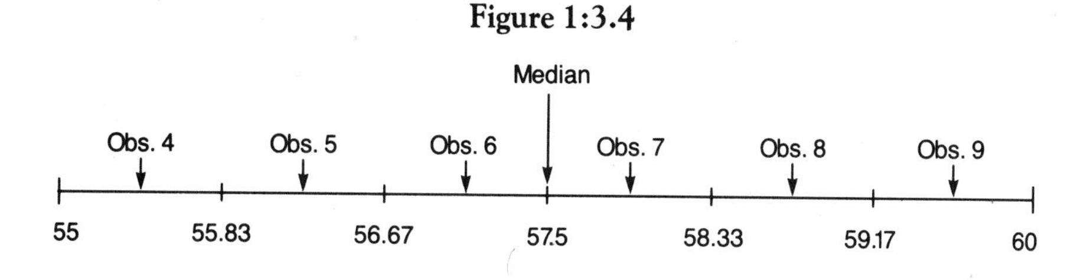

The mean and median are equal to one another for a symmetric histogram. The actual average annual temperature in Reno is 48.4 degrees.

5. mean = [(1,500 × 7) + (2,500 × 3) + (3,500 × 2) + (4,500 × 1)] ÷ 13

   = $2,269.23

   The median for thirteen observations is the seventh observation. It lies in the first class. Subdivide the first class into seven equal subclasses. Assign the seven observations. The median is the seventh observation. Looking at Figure 1:3.5, it is (1,857.13 + 2,000) ÷ 2 = $1,928.56.

Figure 1:3.5

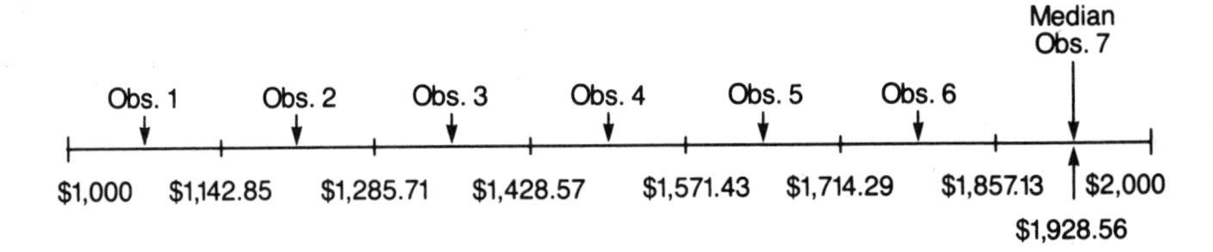

6. It is impossible for everyone to do better than the average. The average is the sum of the observations divided by the sample size. The mean cannot be smaller than the smallest value. The mean must fall somewhere between the smallest and largest value.

7. The histogram data are easier to comprehend. Fifty numbers written on a page are difficult to decipher but one picture organizes the data for us.

Exercise Set Solutions for 1:4

1. range = 14 − 4 = 10

   variance = $[(4-8)^2 + (4-8)^2 + (5-8)^2 + \ldots + (14-8)^2] \div 10 - 1$

   = 15.33

   standard deviation = 3.92

2. The histogram on the left has the greater variance. The mean is at the center of the middle class in each histogram. In the histogram on the left, six of the nine observations are in classes far away from the mean. Only two out of nine observations are in classes far away from the mean in the histogram on the right. Thus the variance is greater for the histogram on the left.

3. Since the variance is the squared dispersion around the mean, it can never be less than zero. It is impossible to have less dispersion than *no* dispersion.

4. College cost data

$$\text{variance} = [(1{,}500 - 2{,}269.23)^2 \times 7 + \ldots + (4{,}500 - 2{,}269.23)^2 \times 1] \div 13 - 1$$

$$= 1{,}025{,}641$$

standard deviation = \$1,012.74

Reno temperature data

$$\text{variance} = [(47.5 - 57.5)^2 \times 1 + \ldots + (67.5 - 57.5)^2 \times 1] \div 12 - 1$$

$$= 27.27$$

standard deviation = 5.22 degrees

Exercise Set Solutions for 1:5

1. College cost data

mean = \$2,269.23

standard deviation = \$1,012.74

In 1972–1973, at least 75 percent of all schools cost between the mean plus or minus two standard deviations. Between 2,269.23 − (2 × 1,012.74) or \$243.75 and 2,269.23 + (2 × 1,012.74) or \$4,294.71.

Reno temperature data

mean = 57.5 degrees

standard deviation = 5.22 degrees

At least 75 percent of all students guessed that the average annual temperature in Reno was between the mean plus or minus two standard deviations. Between 57.5 − (2 × 5.22) or 47.06 degrees and 57.5 + (2 × 5.22) or 67.94 degrees.

2.

$$\text{mean} = 8$$

$$\text{standard deviation} = 3.92$$

Here's one of the "at least" statements. At least 75 percent ($h = 2$) of the observations must fall between the mean plus or minus two standard deviations. This is between .17 and 15.83. In actuality, all ten observations (or 100 percent) fall between these two numbers.

Exercise Set Solutions for 2:2

Problem set solutions for basic probability

1. There are 52 possible outcomes. You are interested in only tens. There are four tens. The probability is $^4/_{52}$.
2. The tree diagram for this problem is shown in Figure 2:2.2. The first draw from the urn can be represented by five branches (R, R,

Figure 2:2.2

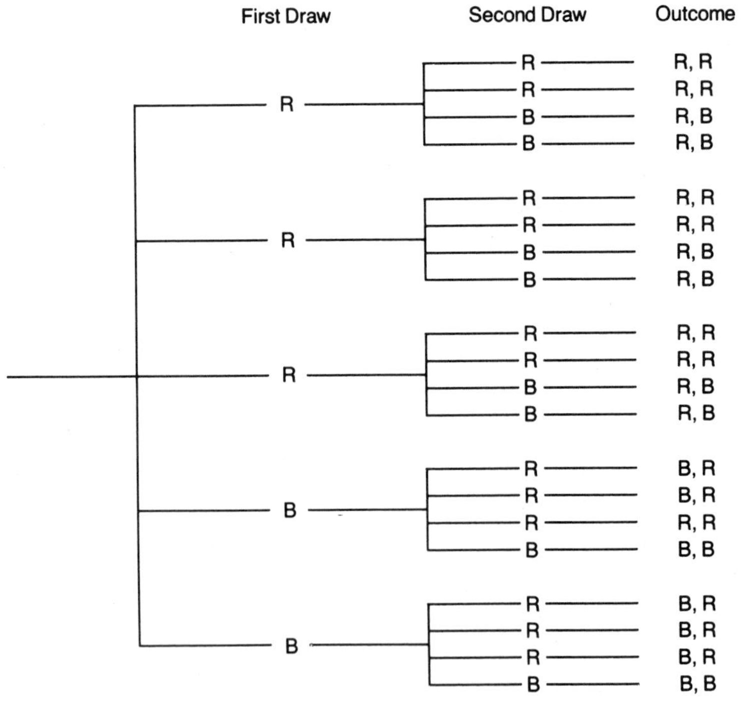

First Draw	Second Draw	Outcome
R	R	R, R
	R	R, R
	B	R, B
	B	R, B
R	R	R, R
	R	R, R
	B	R, B
	B	R, B
R	R	R, R
	R	R, R
	B	R, B
	B	R, B
B	R	B, R
	R	B, R
	R	R, R
	B	B, B
B	R	B, R
	R	B, R
	R	B, R
	B	B, B

R, B, and B). The chances of a red marble on the first draw are three out of five. But after drawing a red marble there are only two red marbles left out of the four remaining marbles. There are five branches at Draw 1 and only four branches at Draw 2. The total number of possible outcomes is 20. We are interested in only six of the 20 branches. The probability of two red marbles is $^6/_{20}$ or .30.

3. The probability will be different if you replace the marble after the first draw. It will get larger. The probability of getting a red marble on the first draw is the same as in the problem above. But if you draw a red marble and then replace it, the probability of getting a red marble on the second draw is greater than in the previous problem. Now it's three out of five—the same as on the first draw. If you drew a tree diagram it should have five branches at the first draw and five branches at the second draw. The total number of possible outcomes in two draws is 25 events. There are nine branches with two red marbles. The probability is $^9/_{25}$ or .36.
4. Each fork of the tree will have two branches, boy or girl. There are four levels of forks to represent four children. The total number of possible outcomes is 16. Only one branch has four boys. The probability of four boys is $^1/_{16}$ or .0625.

Problem set solutions for expected value and standard deviation

1. The probability of no boys is $^1/_{16}$ or.0625.
   The probability of one boy is $^4/_{16}$ or .25.
   The probability of two boys is $^6/_{16}$ or .375.
   The probability of three boys is $^4/_{16}$ or .25.
   The probability of four boys is $^1/_{16}$ or .0625.
   Since these are all the possible events, the sum of their probabilities must be 1.0.

$$\text{expected value} = (0 \times .0625) + (1 \times .25) + (2 \times .375) + (3 \times .25) + (4 \times .0625)$$

$$= 2 \text{ boys in a family with four children}$$

$$\text{variance} = (0 - 2)^2 \times .0625 + (1 - 2)^2 \times .25 + (2 - 2)^2 \times .375 + (3 - 2)^2 \times .25 + (4 - 2)^2 \times .0625$$

$$= 1$$

$$\text{standard deviation} = 1 \text{ boy}$$

2. Expected profit $= 25{,}000 \times 1 + 45{,}000 \times .5 + 60{,}000 \times .3 + 75{,}000 \times .1$

   $= \$50{,}500$

   No. The expected value is a meaningful figure for the long run. If we were faced with the same profits and the same set of probabilities over many quarters, our expected profit would average \$50,500. Only if the standard deviation is small will the expected value be a meaningful figure for the upcoming quarter.

3. You could conclude that you made a math error. Variance or standard deviation can never be less than zero.

   $$\text{variance} = (25{,}000 - 50{,}500)^2 \times .1 + (45{,}000 - 50{,}500)^2 \times .5 + (60{,}000 - 50{,}500)^2 \times .3 + (75{,}000 - 50{,}500)^2 \times .1$$

   standard deviation = \$12,932.52

   As compared to the expected value of \$50,500, the standard deviation is not small. It is over ¼ the value of the expected value.

Exercise Set Solutions for 2:4

1.

	In Favor	Not in Favor	Total
Less than or equal to \$30,000	40	80	120
Greater than \$30,000	50	30	80
Total	90	110	200

The probability of being in favor of a domed stadium given your income is less than or equal to \$30,000 is $^{40}/_{120}$. The probability of being in favor of a domed stadium given your income is greater than \$30,000 is $^{50}/_{80}$. Since the two probabilities are not the same or even close, it means that one's position is influenced by level of income. Income and position appear to be statistically dependent events.

2.

	Democrat	Republican	Independent	Total
In Favor	100	250	100	450
Not in Favor	500	150	100	750
Total	600	400	200	1,200

3. a. The probability of a selected voter being a Democrat is $^{600}/_{1,200}$.
   b. The probability of a voter being in favor of increased spending given he or she is a Republican is $^{250}/_{400}$.
   c. The probability that a voter is a Democrat given that he or she is in favor of increased military spending is $^{100}/_{450}$.

4. Thirty percent of all the workers have been promoted ($^{60}/_{200}$). The conditional probability of being promoted given you are a male and the conditional probability of being promoted given you are a female must also be 30 percent. In that way the person's sex does not affect his or her chance of being promoted. Thirty percent of all males must be promoted. Thirty percent of all females must be promoted. Now you can construct a contingency table.

	Promoted	Not Promoted	Total
Male	36	84	120
Female	24	56	80
Total	60	140	200

Double check: The probability of being promoted given you are a male is $^{36}/_{120}$ or .30. The probability of being promoted given you are a female is $^{24}/_{80}$ or .30.

Exercise Set Solutions for 2:6

1. 100 People

70 Pass		30 Fail	
63 higher than 550	7 not higher than 550	12 higher than 550	18 not higher than 550

	Pass	Fail	Total
Less Than or Equal to 550 SAT	7	18	25
Higher Than 550 SAT	63	12	75
Total	70	30	100

The probability of a student passing given he or she has scored higher than 550 is $^{63}/_{75}$ or .84.

2\.

100 People			
30 Republicans		70 Non-Republicans	
18 in favor	12 not in favor	21 in favor	49 not in favor

	Non-Republican	Republican	Total
In Favor	21	18	39
Not in Favor	49	12	61
Total	70	30	100

The probability of selecting a Republican given the person is in favor of the amendment is $^{18}/_{39}$ or .46.

3\.

100 People			
10 Creative		90 Not Creative	
6 over 70	4 not over 70	4.5 over 70	85.5 not over 70

	Creative	Non-creative	Total
Less Than or Equal to 70	4	85.5	89.5
Higher Than 70	6	4.5	10.5
Total	10	90	100

The probability of selecting a creative person given he or she scored higher than 70 on the test is $^{6}/_{10.5}$ or .57.

4.

1,000 People

10 Hepatitis		990 No Hepatitis	
8.5 tests show Hepatitis	1.5 tests show no Hepatitis	99 tests show Hepatitis	891 tests show no Hepatitis

	Hepatitis	No Hepatitis	Total
Tests Show Hepatitis	8.5	99	107.5
Tests Show No Hepatitis	1.5	891	892.5
Total	10	990	1,000

The probability of having hepatitis given the test says that you have it is $^{8.5}/_{107.5} = .079$.

Exercise Set Solutions for 3:3

1. P(boy and boy and girl) = $.487 \times .487 \times .513 = .1217$
2. P(heart and heart and non-heart) = $^{13}/_{52} \times {}^{13}/_{52} \times {}^{39}/_{52}$ $= .0468$
3. Yes, it would. The chance of getting a heart does not remain constant from one draw to the next. We do not have a Bernoulli process.

   P(heart on Draw 1) × P(heart on Draw 2 given heart on Draw 1) ×

   P(non-heart given two hearts already drawn)

   $= {}^{13}/_{52} \times {}^{12}/_{51} \times {}^{39}/_{50}$

   $= .0458.$

   This answer is slightly different from the answer in Problem 2.
4. The probability that all five ovens work on any given day, including next Tuesday, is

   $$.9 \times .9 \times .9 \times .9 \times .9 = .59$$
5. P(r and r and r and r and r and r and b and b and b and b)

   $$^{70}/_{100} \times {}^{69}/_{99} \times {}^{68}/_{98} \times {}^{67}/_{97} \times {}^{66}/_{96} \times {}^{65}/_{95} \times {}^{30}/_{94} \times {}^{29}/_{93} \times {}^{28}/_{92} \times {}^{27}/_{91} = .000988$$

Exercise Set Solutions for 3:5

1. $$\text{P(no high volume customers)} = \frac{4!}{0! \times 4!} \times (.8 \times .8 \times .8 \times .8)$$
$$= .4096$$

2. No. To use the basic probability expression in Chapter 2, the probability of selecting a low or a high volume purchaser would have to be the same. They do not have to be the same when computing binomial probabilities.

3. $$\text{P(one high volume customer)} = 4! \div (3! \times 1!) \times [.2 \times .8 \times .8 \times .8]$$
$$= .4096$$
$$\text{P(two high volume customers)} = 4! \div (2! \times 2!) \times [.2 \times .2 \times .8 \times .8]$$
$$= .1536$$
$$\text{P(three high volume customers)} = 4! \div (1! \times 3!) \times [.2 \times .2 \times .2 \times .8]$$
$$= .0256$$
$$\text{P(four high volume customers)} = 4! \div (0! \times 4!) \times [.2 \times .2 \times .2 \times .2]$$
$$= .0016$$

4. Figure 3:5.4 is the histogram for the probability distribution of the number of high volume purchasers in a sample of four customers.

Figure 3:5.4

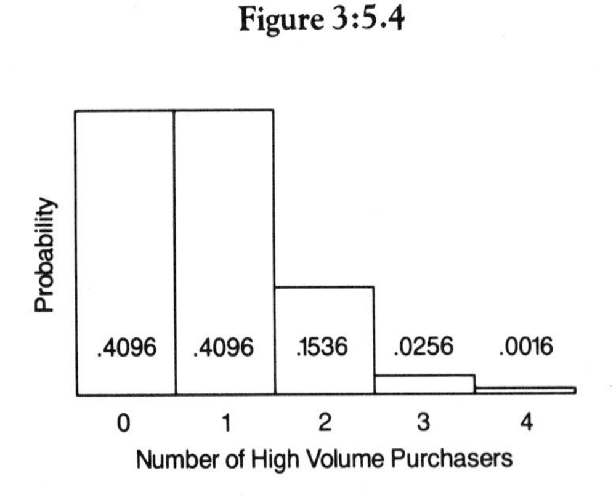

Figure 3:5.7

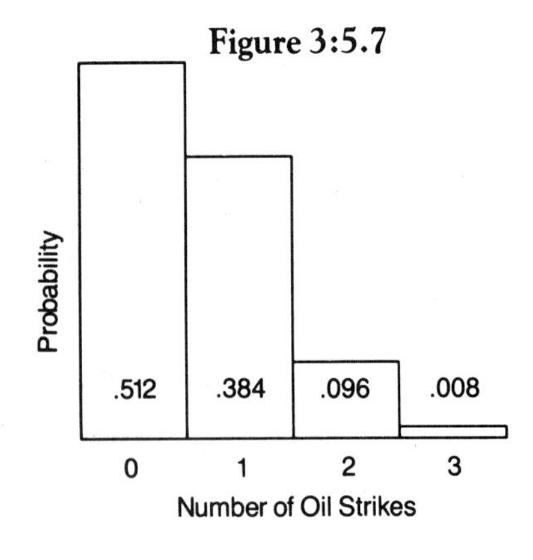

5. $$\text{expected value} = (0 \times .4096) + (1 \times .4096) + (2 \times .1536) + (3 \times .0256) + (4 \times .0016)$$
$$= .8 \text{ high volume customers}$$
$$\text{variance} = (0 - .8)^2 \times .4096 + (1 - .8)^2 \times .4096 + (2 - .8)^2 \times .1536 + (3 - .8)^2 \times .0256 + (4 - .8)^2 \times .0016$$
$$= .64$$
standard deviation = .8 high volume charge customers

6. $$\text{P(no oil strikes)} = 3! \div (0! \times 3!) \times [.8 \times .8 \times .8]$$
$$= .512$$
$$\text{P(one oil strike)} = 3! \div (1! \times 2!) \times [.2 \times .8 \times .8]$$
$$= .384$$
$$\text{P(two oil strikes)} = 3! \div (2! \times 1!) \times [.2 \times .2 \times .8]$$
$$= .096$$
$$\text{P(three oil strikes)} = 3! \div (3! \times 0!) \times [.2 \times .2 \times .2]$$
$$= .008$$

7. Figure 3:5.7 on the preceding page is the probability histogram for the number of oil strikes out of three wells.

8. $$\text{expected value} = 0 \times .512 + 1 \times .384 + 2 \times .096 + 3 \times .008$$
$$= .6 \text{ oil wells}$$
$$\text{variance} = (0 - .6)^2 \times .512 + (1 - .6)^2 \times .384 + (2 - .6)^2 \times .096 + (3 - .6)^2 \times .008$$
$$= .48$$
standard deviation = .69 oil wells

Exercise Set Solutions for 3:6

1. To compute the probability of a bottle containing less than or equal to 15.7 ounces, convert 15.7 ounces to a z-score.

$$z\text{-score} = (15.7 - 16) \div .3 = -1 \text{ standard units}$$

The probability is

$$.5 - .3413 = .1587$$

2. To compute the probability of a bottle containing between 15.85 and 16.60 ounces, convert 15.85 to a z-score.

$$z\text{-score} = (15.85 - 16) \div .3 = -.5 \text{ standard units}$$

Convert 16.60 ounces to a z-score.

$$z\text{-score} = (16.60 - 16) \div .3 = 2 \text{ standard units}$$

The probability is

$$.1915 + .4772 = .6687$$

3. The probability of a bottle containing 16.09876859402345 is zero because there is no area under the normal curve at a single point.

4. The probability of finishing within 15 minutes using the mop:

$$z\text{-score} = (15 - 10) \div 2.5 = 2 \text{ standard units}$$

The probability that the time it takes you to finish will be less than or equal to 2 standard units above the mean is

$$.50 + .4772 = .9772$$

Note: The .50 is the probability of finishing in less than the mean time of ten minutes. If you drew the picture, this is the area under the curve to the left of the mean.

The probability of finishing within 15 minutes using the knees/floor method is

$$z\text{-score} = (15 - 14) \div 1 = 1 \text{ standard unit}$$

The probability of finishing in less than or equal to one standard unit is

$$.50 + .3413 = .8413$$

Use the mop and enjoy your soap opera.

Exercise Set Solutions for 4:2

1. Argument (a) is an example of inductive inference. The wedge is a sample of the melon. You draw a general conclusion about the entire melon from the sample.

Argument (b) is an example of deductive inference. The first statement is a general statement about the population of melons. From this statement you draw a specific conclusion about one melon.

2.

	Population	Parameter	Statistic
a.	all beans in the jar	percentage of red beans	$^{20}/_{50} = .40$
b.	families of four within the city	average income of families of four	$20,100
c.	all TV viewers	percentage watching a show	$^{100}/_{1,500} = .067$

3. We would still exclude families who have just moved and have either not obtained their telephone yet or whose names are not yet in the telephone book. We would also eliminate all those with unlisted numbers. We would also eliminate those who still cannot afford a telephone.

4. Picture A is an example of low population standard deviation and high bias. All the shots are close together (low standard deviation) but are not near the target center or the population parameter (high bias).

   Picture B is an example of high population standard deviation and low bias. The shots are not close together but their average would be close to the target center.

   Picture C is an example of low population standard deviation and low bias. The shots are all tightly clustered near the target center.

   Picture D is an example of high population standard deviation and high bias. The shots are not close together and their average is far away from the target center.

5. Yes, because they have systematically excluded people who rent apartments or homes. Since these people are potential customers and they have been eliminated from the sample, we have selection bias.

6. No two machines or people are exactly alike. Even a single individual or machine may not perform exactly the same over time. All of these small differences have an impact on the variability in ring diameters. These include the following human and technological factors.

Human	Technological
Worker differences such as training, motivation, and skill level	Machine differences
Worker fatigue throughout the day	Differences in raw materials
Human error	Working conditions—humidity or temperature

Exercise Set Solutions for 4:4

1. The alternative is to sample the entire population. This is too costly and time-consuming, therefore it is impractical.
2. The standard error of the sampling distribution, $\sqrt{\text{sigma}^2/n}$, decreases as the sample size increases. As the standard error decreases, the chances increase that a given sample mean will be closer to the unknown population mean.
3. Never. The standard error must be less than or equal to the population standard deviation. To compute the standard error, you divide the population standard deviation by the square root of the sample size, $\sqrt{\text{sigma}^2/n}$ or $\text{sigma}/\sqrt{n}$. Thus for sample sizes greater than one, the standard error of the sampling distribution must be smaller than the population standard deviation.
4. The sample mean grade point average, *x*-bar, is 2.8. The sample variance, s^2, is

   $$[(2.25 - 2.8)^2 + \ldots + (3.10 - 2.8)^2] \div (10 - 1) = .256$$

   The sample variance is an estimate of the unknown population variance. The *estimated* standard error of the sampling distribution is

   $$\sqrt{.256/10} = .16$$

   The 90 percent confidence interval is

   2.8 plus or minus t(90 percent, 9df) $\times$ .16

   $2.8 - (1.833 \times .16)$ to $2.8 + (1.833 \times .16)$

   or 2.51 to 3.09

We are 90 percent confident that the unknown population mean grade point average at the university is between 2.51 and 3.09.

5. The sample mean, *x*-bar, is 15 miles.

 The sample variance, s^2, is

$$[(13 - 15)^2 + . . . + (13 - 15)^2] \div (17 - 1) = 33.625$$

The *estimated* standard error is $\sqrt{33.625/17}$ or 1.406 miles. The 99 percent confidence interval is 15 ± t(99 percent, 16df) × 1.406

$$15 - (2.921 \times 1.406) \text{ to } 15 + (2.921 \times 1.406)$$

or 10.89 to 19.11

We are 99 percent confident that the unknown population mean commuting distance is between 10.89 and 19.11 miles.

6. We already know the sample mean; we computed it from the data. We use the sample mean to estimate, with some degree of certainty, the unknown population mean. We *know* sample statistics and we *estimate* unknown population parameters.

7.

$$x\text{-bar} = 10 \text{ inches } s^2 = 144$$

The *estimated* standard error is $\sqrt{144/1{,}000}$ = .379 inches.

 Because of the large sample size, you may use the normal table value of 1.96.

$$10 - (1.96 \times .379) \text{ to } 10 + (1.96 \times .379)$$

or 9.26 to 10.74

We are 95 percent confident that the unknown mean number of inches of insulation for the utility company's customers is between 9.26 and 10.74 inches.

8. The sample proportion, *p*-hat, is $260/1{,}000$ = .26

 The *estimated* standard error of *p*-hat = $\sqrt{(.26 \times .74)/1{,}000}$ = .01387.

 Since the sample size is so large the sampling distribution is approximately normal. We use the *z*-value from the normal table which is 1.96. The interval is

$$.26 - (1.96 \times .01387) \text{ to } .26 + (1.96 \times .01387)$$

or .2328 to .2872

 We are 95 percent confident that the unknown population proportion of customers who have storm windows is between 23.28 percent and 28.72 percent.

9. The sample proportion, p-hat is $^{50}/_{900} = .0556$.

   The *estimated* standard error is $\sqrt{(.0556 \times .9454)/900} = .0076$.

   Again you can use the normal table because the large sample size insures that the sampling distribution is approximately normal. The z-value for a 95 percent confidence level is 1.96.

   $$.0556 - 1.96 \times .0076 \text{ to } .0556 + 1.96 \times .0076$$

   or .0407 to .0705.

   We are 95 percent confident that the unknown proportion of recent graduates who found their statistics course interesting is between 4.07 percent and 7.05 percent.

10. The sample proportion, p-hat, is $^{300}/_{400} = .75$.

    The *estimated* standard error of p-hat $= \sqrt{(.75 \times .25)/400} = .0217$.

    Again, use the normal table value, which is 1.645 for a 90 percent confidence level.

    $.75 - 1.645 \times .0217$ to $.75 + 1.645 \times .0217$ or .714 to .786.

    We are 90 percent confident that the proportion of ACR owners who have had no problems is between 71.4 percent and 78.6 percent.

Exercise Set Solutions for 4:6

1. Null Hypothesis: The long-run average braking distance is greater than or equal to 60 feet.

   Null Action: We will not switch to the new brake system as it is no better than our present brake system.

   Alternative Hypothesis: The long-run average braking distance is less than 60 feet.

   Alternative Action: We will switch to the new brake system.

   Type 1 Error: Reject the null when it is true.
   You switch to the new brake system when in fact it is no better than your present brake system.

Type 1 Error Costs:	New equipment, retooling, additional training costs of workers, removal of equipment to manufacture old brake system, and possible class action legal suits against your company for making false claims about its brakes.
Type 2 Error:	Accept the null when it is false. You fail to switch to the new brake system when it really is better than your present brake system.
Type 2 Error Costs:	Possible loss of sales to a safety conscious buying public not purchasing your cars.

With a sample size of 100, the sampling distribution will be normal. It is drawn in Figure 4:6.1, under the assumption that the null hypothesis is true. The critical value is

$$60 - 1.282 \times (10 \div \sqrt{100}) = 58.7 \text{ feet}$$

Figure 4:6.1

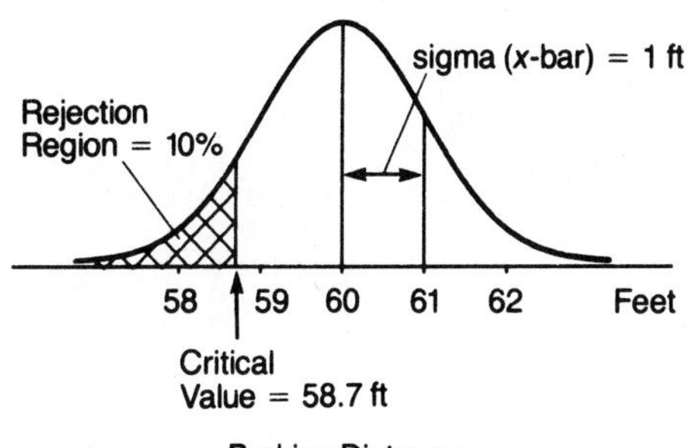

Figure 4:6.2

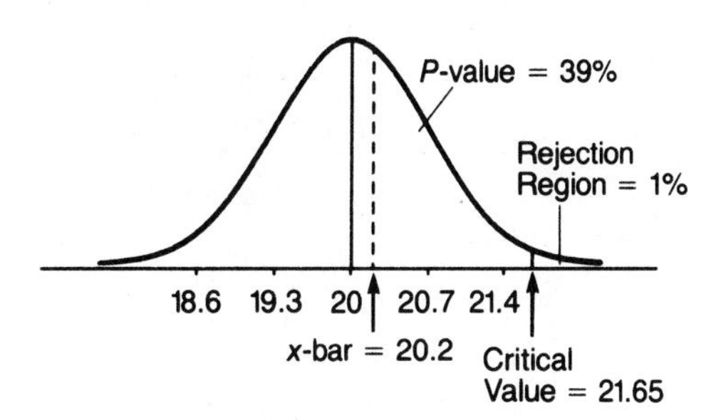

If the sample mean braking distance is greater than or equal to 58.7 feet, accept the null hypothesis. If the sample mean braking distance is less than 58.7 feet, reject the null hypothesis. Since the sample mean braking distance based upon a sample size of 100 is 58 feet we should reject the null hypothesis and switch over to the new brake system.

2. Null Hypothesis:	The long-run average number of good welds produced by the robots is less than or equal to 20 per minute.
Null Action:	Do not purchase the robots.

Alternative Hypothesis:	The long-run average number of good welds produced by the robots is greater than 20 per minute.
Alternative Action:	Purchase the robots.
Type 1 Error:	Reject the null when it is true. You purchase the robots when they will not produce more than 20 good welds per minute.
Type 1 Error Costs:	Cost of robots, installation cost within the plant, etc.
Type 2 Error:	Accept the null when it is false. You don't purchase the robots when you should have. They will produce more than 20 good welds per minute.
Type 2 Error Costs:	Costs associated with losing your competitive advantage. Your costs could escalate due to the failure to purchase the robots. You could lose a share of your market.

The sampling distribution, assuming the null hypothesis is true, is shown in Figure 4:6.2. The critical value is

$$20 + (2.33 \times 7.07 \div \sqrt{100}) = 21.65 \text{ good welds per minute}$$

If the sample average number of good welds is less than or equal to 21.65 per minute, accept the null hypothesis. If the sample average number of good welds is greater than 21.65 per minute, reject the null hypothesis. Since the sample average number of good welds is 20.2, accept the null hypothesis and don't purchase the robots.

3. The level of significance is set prior to the collection of data. It is based upon balancing the costs of making the Type 1 and Type 2 errors. The higher the costs associated with making a Type 1 error, the lower you should set the level of significance. In the previous problem a 1 percent level of significance was set.

   Assume the null hypothesis is true and the population mean is really 20 good welds per minute. The p-value tells you the likelihood of obtaining a sample mean as far away or farther from the hypothesized population mean of 20 as the one you actually obtained. Look at the sampling distribution for Problem 2 in Figure 4:6.2 on page 283. The p-value is equal to the area under the curve to the right of the sample mean of 20.2. This is only .28 standard units from the mean, and the area to the right of it, the *p*-value, is 39 percent. This means that if the null hypothesis is true, 39 percent of the time we would get a sample mean of 20.2 or higher. If we reject the null on the basis of this sample, there is a 39 percent chance that we will have made a Type 1 error. This is a much greater chance than the 1 percent that we were willing to accept. So we do not reject the null.

4. No. Once you make a decision you cannot make both errors. If you accept the null, then you either made a correct decision or you made a Type 2 error. If you reject the null hypothesis, then you either made a correct decision or a Type 1 error. In order to make a Type 1 error you must reject the null when it is true. If you accept the null hypothesis, then it is impossible to make a Type 1 error.

5.

Null Hypothesis:	The person is innocent.
Null Action:	Set the person free.
Alternative Hypothesis:	The person is guilty.
Alternative Action:	Jail the person.
Type 1 Error:	You decide the person is guilty when he or she is really innocent, sending an innocent person to jail.
Type 2 Error:	You decide the person is innocent when he or she is really guilty, freeing a guilty person.

6. The sample average of 20.2 was only based upon a sample of 100 observations. Thus even if the long-run average were less than 20 good welds per minute it is possible for one sample of size 100 to be greater than 20. We need to allow for a margin of error. Therefore, we concluded that we could not say with 99 percent confidence that the long-run average number of good welds was greater than 20 per minute unless the sample average was greater than 21.65 per minute.

7. H_0: Ten percent or more customers are dissatisfied.
H_1: Fewer than 10 percent are dissatisfied.
Assuming the null hypothesis is true, the mean of the sampling distribution is .10 and the standard error is

$$\sqrt{(.1 \times .9) \div 200} = .021$$

The z value for a 10 percent level of significance is 1.282. The critical value is

$$.10 - (1.282 \times .021) = .073$$

The rejection region is below .073. Since the sample proportion is .04, the firm rejects the null. They are at least 90 percent confident that the new program is working.

Exercise Set Solutions for 5:2

1. **ANOVA Table**

Sources of Variation	Sum of Squares	df	Variance	Variance Ratio
Treatment	24	2	12	18
Extraneous	4	6	.67	
Total	28	8		

Given a variance ratio of 18, you are at least 99 percent confident [F(2, 6, 99 percent) = 10.92] that the null hypothesis is false and that the different merchandising displays affect sales revenue. You should reject the null hypothesis.

2. Since variances can never be negative, you must have made a mathematical error. Go back and check your work.

3. **ANOVA Table**

Sources of Variation	Sum of Squares	df	Variance	Variance Ratio
Treatment	24	2	12	18
Extraneous	6	9	.67	
Total	30	11		

Given a variance ratio of 18, you are at least 99 percent confident [F(2, 9, 99 percent) = 8.02] that the null hypothesis is false and that the different drugs have an impact on weight loss. You should reject the null hypothesis.

4. You must use randomization to accomplish the "RO" principle. Twelve underweight subjects would be selected from the population and randomly divided into three groups of four subjects each. Each group would be assigned to one of the drugs at random. If the entire experiment could not be carried out simultaneously, the sequence by which subjects receive their drug would be determined using a random numbers table.

5. **ANOVA Table**

Sources of Variation	Sum of Squares	df	Variance	Variance Ratio
Treatment	14	1	14	21
Extraneous	8	12	.67	
Total	22	13		

Yes, you are more than 99 percent confident [F(1, 12, 99 percent) = 9.33] that the null hypothesis is false and that color displays improve one's understanding of the information.

6. We draw an inductive inference when we compare the variance ratio from the experimental data with the table's Fisher values. If the variance ratio is larger than the F value, we reject the no difference null hypothesis at the associated level of confidence; otherwise we don't. We draw inductive inferences because we only take a sample from a population and need to draw inferences back to the population. We are interested in long-run behavior and small samples do not constitute long-run behavior.

Exercise Set Solutions for 5:3

1. The profiles for this study are shown in Figure 5:3.1.

Figure 5:3.1

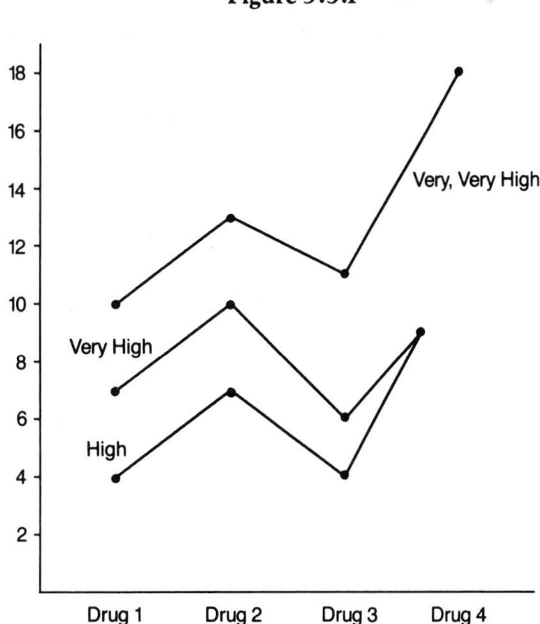

ANOVA Table

Sources of Variation	Sum of Squares	df	Variance	Variance Ratio
Block	104	2		
Treatment	54	3	18	9
Extraneous	12	6	2	
Total	170	11		

You are at least 95 percent confident [F(3, 6, 95 percent) = 4.76] that the null hypothesis is false. However, you cannot be 99 percent confident that the null is false [F(3, 6, 99 percent) = 9.78]. If the null hypothesis is true, there is a greater than 1 percent probability that you could have gotten these results by chance.

2. Patients with high blood pressure are subdivided into three groups depending on their systolic blood pressure—high, very high, and very, very high. Three subjects are randomly selected from each classification. Each subject in a classification is randomly assigned to a drug. In actual experiments like this, the drugs are coded so the patients do not know which drugs they receive and the doctors do not know which drug each patient receives. This is called a "double blind" experiment.

3. The profiles for the computer software study are shown in Figure 5:3.3.

Figure 5:3.3 **Figure 5:3.4**

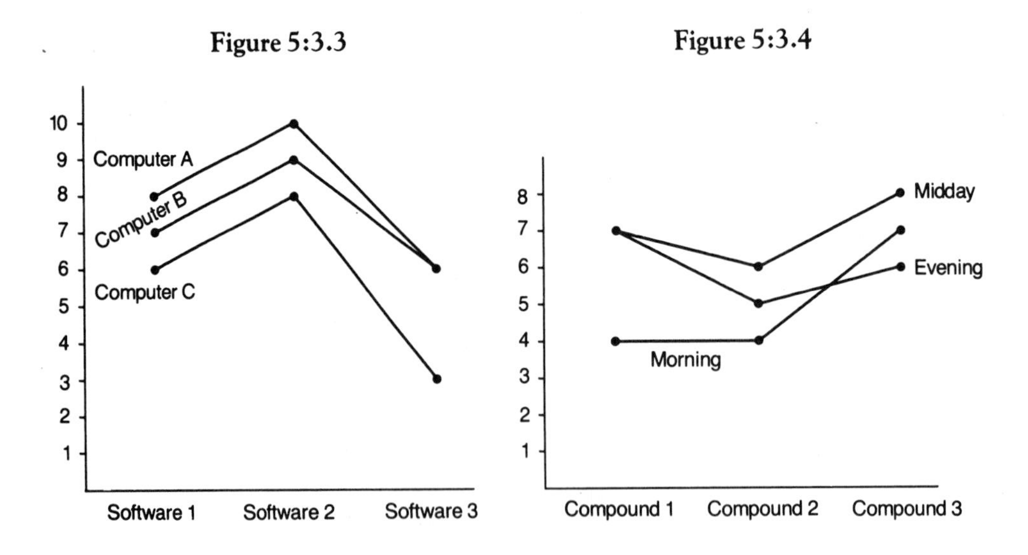

ANOVA Table

Source of Variation	Sum of Squares	df	Variance	Variance Ratio
Block	8.67	2		
Treatment	24	2	12	36
Extraneous	1.33	4	.33	
Total	34	8		

You are at least 99 percent confident [F(2, 4, 99 percent) = 18] that the no difference in run times null hypothesis is false.

4. The profiles for this problem are shown in Figure 5:3.4 on the preceding page.

ANOVA Table

Sources of Variation	Sum of Squares	df	Variance	Variance Ratio
Block	6	2		
Treatment	6	2	3	3
Extraneous	4	4	1	
Total	16	8		

You are much less than 90 percent confident [F(2, 4, 90 percent) = 4.32] that the null hypothesis is false; that is, you are not very confident that the null is false. You cannot conclude that the three compounds have different effects on miles per gallon.

Exercise Set Solutions for 5:4

1. Before plotting the need for independence profiles, compute the four cell means: 12, 16, 16, and 8. The profiles are shown in Figure 5:4.1 on the next page.

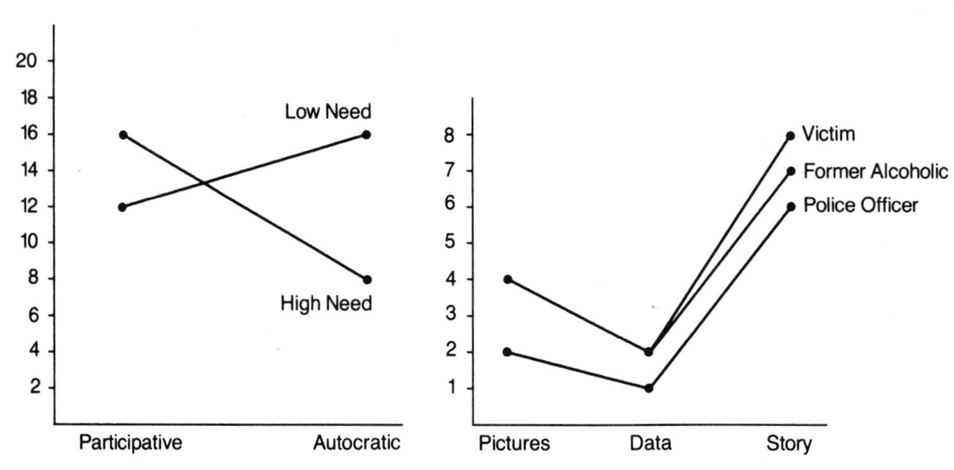

ANOVA Table

Sources of Variation	Sum of Squares	df	Variance	Variance Ratio
Treatment A	8	1	8	
Treatment B	8	1	8	
AB Interaction	72	1	72	7.2
Extraneous	40	4	10	
Total	128	7		

You can be more than 90 percent confident [F(1, 4, 90 percent) = 4.54] but less than 95 percent confident [F(1, 4, 95 percent) = 7.71] that there is a long-run interaction effect between need for independence and style of supervision on worker productivity.

2. Workers with low need for independence are best led by an autocratic leader. Workers with a high need for independence are best led by a participative leader. The best leadership style depends upon the need for independence of the workers.

3. Before drawing Factor A profiles, compute the averages for the nine cell means: 4, 2, 7, 4, 2, 8, 2, 1, and 6. The Factor A profiles are shown above in Figure 5:4.3.

ANOVA Table

Sources of Variation	Sum of Squares	df	Variance	Variance Ratio
Factor A	9.36	2	4.68	5.26
Factor B	89.28	2	44.64	50.16
AB Interaction	1.36	4	.34	.38
Extraneous	8	9	.89	
Total	108	17		

The variance ratio for the AB interaction is less than 1. This means that the variance due to the interaction effect is less than the variance due to the random influence of extraneous factors. You should conclude that you cannot detect an interaction between the two experimental factors. Now we test the main effects. You are more than 95 percent confident [F(2, 9, 95 percent) = 4.26] that the null hypothesis for Factor A, that there is no difference among presenters, is false. You are more than 99 percent confident [F(2, 9, 99 percent) = 8.02] that the null hypothesis for Factor B is false. The type of presenter and the type of presentation do affect students' attitudes towards starting a SADD chapter. It appears, subject to formal verification using confidence intervals, that the police officer is the least effective presenter.

It appears, subject to formal verification using confidence intervals, that the individual story is the most effective type of presentation.

To get students to start a SADD chapter, either a former alcoholic or the victim ought to tell a story about a particular DUI accident. This combination appears to be the most effective overall.

4. Since both experimental factors refer to differences in the presentation, not the subjects, all 18 students should be drawn from the same population. Subjects should be selected randomly from the entire student body and randomly assigned to fill the nine experimental cells.

Exercise Set Solutions for 6:4

1. The scatter diagram, which is shown in Figure 6:4.1 on the next page, can be encircled by a tight ellipse sloping upward to the right. Thus it appears that the variables are linearly and positively related.

Figure 6:4.1

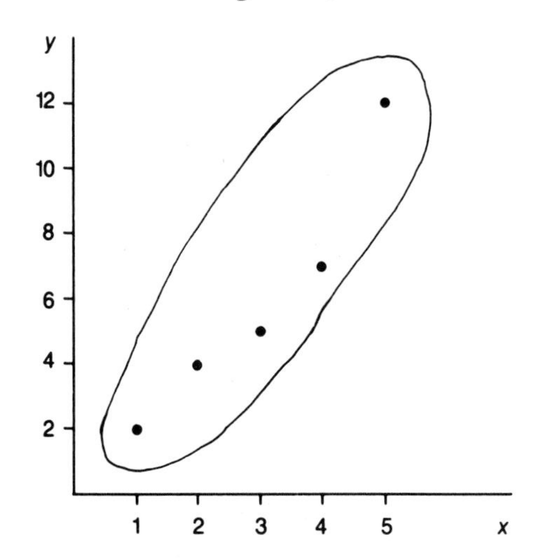

Figure 6:4.4

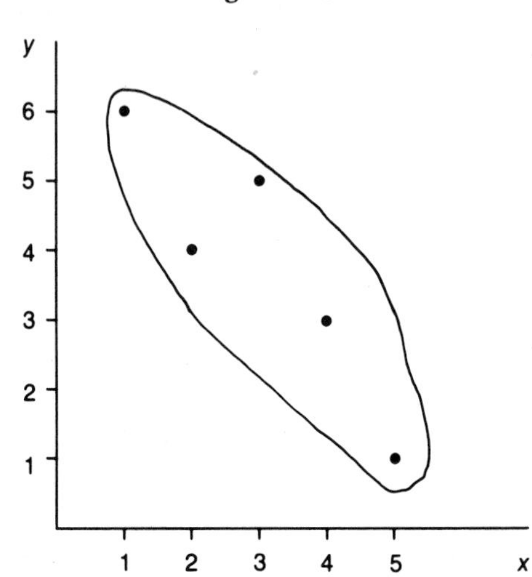

$$b = \frac{113 - (5 \times 3 \times 6)}{55 - (5 \times 3 \times 3)}$$

$$= 2.3$$

$$a = 6 - (3 \times 2.3)$$

$$= -.9$$

$$y = -.9 + 2.3x$$

The sum of squared deviations = 5.10.

2. The scatter diagram for this problem is shown in Figure 6-4 on page 236. The scatter diagram cannot be encircled by an ellipse; it will take a circle to encompass the data. It does not appear that there is a linear relationship between the two variables. Just by observation it appears that the data are nonlinearly related. As x gets bigger, first y increases and then y decreases.

$$b = \frac{39 - (5 \times 3 \times 2.6)}{55 - (5 \times 3 \times 3)}$$

$$= 0$$

$$a = 2.6 - (0 \times 3)$$

$$= 2.6$$

$$y = 2.6 + 0x$$

The sum of squared deviations is 11.2.

3. Another way to eliminate the positive deviations cancelling out the negative deviations is to take the absolute values. This approach calls for minimizing the sum of absolute values of the deviations.

4. It does appear that the two variables are linearly related since all the data can be encompassed by a tight ellipse that slopes downward to the right, as shown in Figure 6:4.4 on the preceding page.

$$b = \frac{46 - (5 \times 3 \times 3.8)}{55 - (5 \times 3 \times 3)}$$

$$= -1.1$$

$$a = 3.8 - (-1.1 \times 3)$$

$$= 7.1$$

$$y = 7.1 - 1.1x$$

The sum of squared deviations is 2.70.

5. a. The sum of squares can never be less than zero since we square the deviations.

   b. The smallest possible value is zero.

   c. When the sum of the squared deviations is zero, the best fitting linear line must pass through all the data points.

Exercise Set Solutions for 6:5

1. The scatter diagram for this problem is shown in Figure 6:5.1 on the next page.

$$b = \frac{1{,}854 - (6 \times 5 \times 66.667)}{220 - (6 \times 5 \times 5)}$$

$$= -2.086$$

$$a = 66.67 - (-2.086 \times 5)$$

$$= 77.10$$

$$y = 77.10 - 2.086x$$

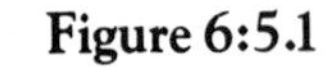

Figure 6:5.1

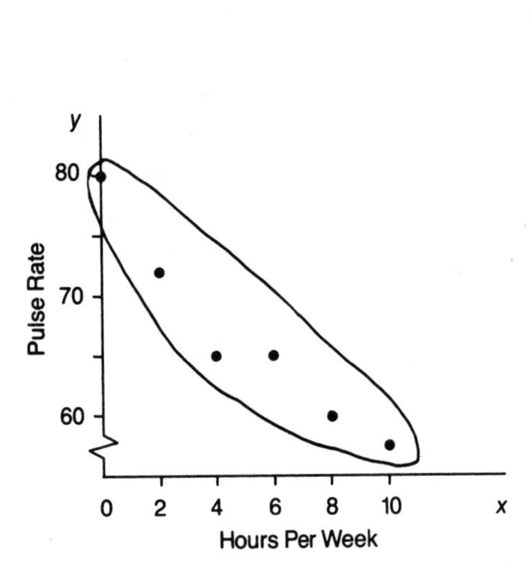

Figure 6:5.2

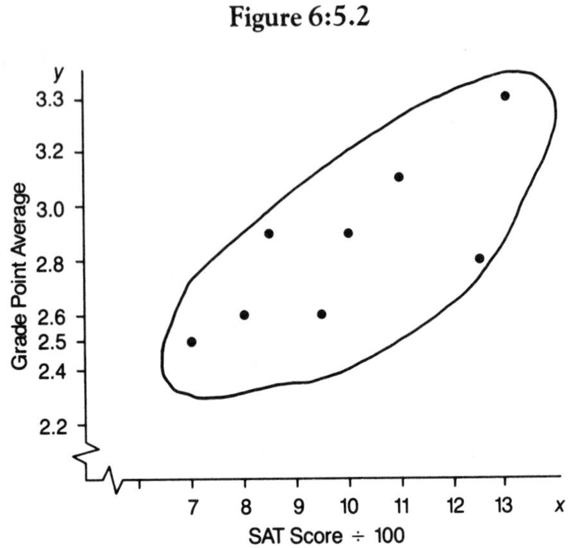

ANOVA Table

Sources of Variation	Sum of Squares	df	Variance	Variance Ratio
Treatment	304.1	1	304.1	44.7
Extraneous	27.2	4	6.8	
Total	331.3	5		

You should be much more than 99 percent confident [F(1, 4, 99 percent) = 21.20] that the no linear association hypothesis is false. The two variables appear to be negatively related ($b = -2.086$).

2. The scatter diagram for the SAT score–grade point average study is shown above in Figure 6:5.2.

$$b = \frac{228.65 - (8 \times 9.9375 \times 2.8375)}{821.75 - (8 \times 9.9375 \times 9.9375)}$$

$$= .0967$$

$$a = 2.8375 - (.0967 \times 9.9375)$$

$$= 1.876$$

$$y = 1.876 + .0967x$$

ANOVA Table

Sources of Variation	Sum of Squares	df	Variance	Variance Ratio
Treatment	.297	1	.297	8.03
Extraneous	.222	6	.037	
Total	.519	7		

You can be more than 95 percent confident [F(1, 6, 95 percent) = 5.99] but less than 99 percent confident [F(1, 6, 99 percent) = 13.74] that the no linear association null hypothesis is false. There appears to be a positive ($b = .0967$) relationship.

3. There are no other factors beyond the treatment or independent variable that affect the dependent variable.

4. Figure 6:5.4 on the next page is the scatter diagram of the life insurance survey results.

$$b = \frac{28{,}660 - (6 \times 50.3333 \times 71.6667)}{18{,}950 - (6 \times 50.3333 \times 50.3333)}$$

$$= 1.871$$

$$a = 71.6667 - (1.871 \times 50.3333)$$

$$= -22.51$$

$$y = -22.51 + 1.871x$$

ANOVA Table

Source of Variation	Sum of Squares	df	Variance	Variance Ratio
Treatment	13,131.30	1	13,131.30	52.42
Extraneous	1,002.04	4	250.51	
Total	14,133.33	5		

You can be at least 99 percent confident [F(1, 4, 99 percent) = 21.20] that the no linear association hypothesis is false.

5. The scatter diagram for this problem is shown in Figure 6:5.5.

Figure 6:5.4

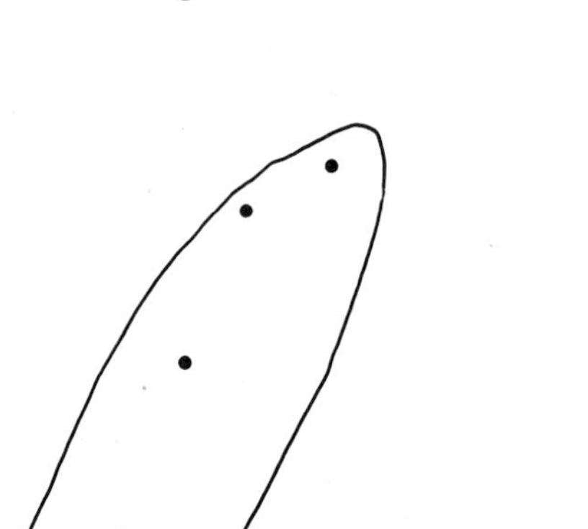

Figure 6:5.5

$$b = \frac{3{,}250 - (6 \times 78.3333 \times 6.833)}{38{,}350 - (6 \times 78.3333 \times 78.3333)}$$

$$= .025$$

$$a = 6.833 - .025 \times 78.3333$$

$$= 4.875$$

$$y = 4.875 + .025x$$

ANOVA Table

Source of Variation	Sum of Squares	df	Variance	Variance Ratio
Treatment	0.958	1	0.958	.214
Extraneous	17.875	4	4.469	
Total	18.833	5		

You are very much less than 90 percent confident [F(1, 4, 90 percent) = 4.54] that the null hypothesis is false. Since the variance ratio is less than 1 you know that the extraneous factors accounted for more of the variation in performance ratings than communication scores did. You should conclude that you cannot detect a relationship between communication score and job performance.

Exercise Set Solutions for 6:6

1.

$$\text{MLV} = 77.10 - (2.086 \times 7) = 62.50$$

$$t(99 \text{ percent, } 4 \text{ df}) = 4.604$$

$$\text{Variance due to extraneous factors} = 6.8$$

$$n = 6$$

The estimated standard error of the most likely value is

$$\sqrt{6.8\left(1 + \frac{1}{6} + \frac{4}{70}\right)} = 2.88$$

The 99 percent prediction interval on y for $x = 7$ hours of exercise per week is

$$\text{MLV} \pm (t \times \text{standard error of MLV})$$

$$62.50 \pm (4.604 \times 2.88)$$

The lower limit is $62.50 - 13.26 = 49.24$
The upper limit is $62.50 + 13.26 = 75.76$

We are 99 percent confident that an individual who exercises seven hours a week should have a pulse rate between 49.2 and 75.8 beats per minute.

2. Pulling apart the values of x can often be done without increasing the cost of the regression study. Increasing the sample size or adding independent variables will increase the cost of collecting the data.

3.

$$\text{MLV} = -22.51 + (1.871 \times 80) = 127.17$$

$$\text{Estimated standard error of MLV} = 18.74$$

$$t(95 \text{ percent, } 4 \text{ df}) = 2.776$$

The 95 percent prediction interval is

$$127.17 \pm (2.776 \times 18.74)$$

The lower limit is $127.17 - 52.02 = 75.15$
The upper limit is $127.17 + 52.02 = 179.19$

We are 95 percent confident that an individual with \$80,000 income will have between \$75,150 and \$179,190 of whole life insurance.

4. The best fitting linear line and the prediction intervals are drawn in Figure 6:6.4. The three prediction intervals are

For 0 hours: 77.10 ± 6.87, or 70.23 to 83.97
For 5 hours: 66.67 ± 6.01, or 60.66 to 72.68
For 10 hours: 56.24 ± 6.87, or 49.37 to 63.11

Figure 6:6.4

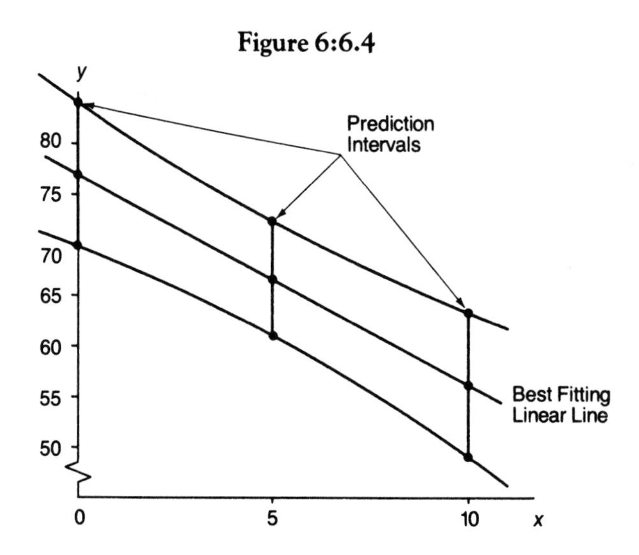

The upper and lower prediction interval limits are curved lines. The prediction interval is narrowest when making predictions for the average value of x. The prediction intervals become wider for values of x above or below the mean or average value of x.

This tells us that the narrowest, and therefore the most reliable, predictions are made for values of x near the mean value of x. As you make predictions for x which are very far away from the mean, your prediction intervals become wider and therefore less meaningful. This should warn you that making extrapolating predictions is risky business.

Index

A

B

D

E

F

G

H

I

J

K

L

M

N

O

P

Q

R

S

T

W

Z